Microbial Biofilms

Current Research and Practical Implications

Edited by

Arindam Mitra
Department of Microbiology
School of Life Science and Biotechnology
Adamas University
Kolkata – 700126
West Bengal, India

Caister Academic Press

www.caister.com

Caister Academic Press, UK
www.caister.com

ISBN: 978-1-912530-32-8 (paperback)
ISBN: 978-1-912530-33-5 (ebook)

DOI: https://doi.org/10.21775/9781912530328

Contents

Current books of interest

Preface

Biofilms are classic examples of microbial communities that persist collectively in a self-synthesized matrix and challenge the concept of prokaryotes as isolated organisms. Microbial biofilms are extremely robust in terms of resistance to various chemicals and antimicrobials and are relevant in more than half of infectious diseases globally. In addition microbial biofilms have numerous industrial applications such as their use in bioremediation, electricity generation and wastewater treatment. The importance of biofilms cannot be understated and work in this field will continue to grow in the future.

This book covers various aspects of microbial biofilms with a focus on mechanisms of biofilm formation, techniques to study microbial biofilms and applications of biofilms. The book is up-to-date in terms of content and developments in the field. Keeping the focus on basics, the book touches on relevant updates in the field of biofilms.

The authors aim to provide a broader overview covering basic to applied research on biofilms. This volume will be useful to graduate students, researchers and scientists working in this field. The book is also relevant to researchers and students from other disciplines wishing to learn more about biofilms.

Arindam Mitra, Ph.D
Associate Professor and Head,
Department of Microbiology
School of Life Science and Biotechnology
Adamas University

Acknowledgements

I would like to thank all authors for their valuable time, effort, energy and patience in contributing towards this concise volume on microbial biofilms. I would also like to thank my doctoral mentor, Dr. Suman Mukhopadhyay who introduced me to the exciting world of biofilms for the first time. My sincere thanks to respected Prof. Samit Ray, Chancellor, Adamas University and respected Prof. Madhusudan Chakraborty, Vice Chancellor, Adamas University for their support over the years at Adamas University.

This volume would not be possible without the patience and cooperation of Hugh Griffin throughout the publishing process of this book. I would also like to take this opportunity to humbly thank all scientists and investigators who keeps on working tirelessly to find new ways to explore biofilms.

Finally, I can't thank enough my better half, Apabrita for her constant support and belief in me. My gratitude goes to all my family members and friends for sticking together through thick and thin akin to microbes in biofilms.

Arindam Mitra, Ph.D
Associate Professor
Department of Microbiology
School of Life Science and Biotechnology
Adamas University
Kolkata, India

Molecular Mechanisms of Biofilm Development and Biofilm Dispersal in Gram-positive Bacteria

1

Öykü İrigül-Sönmez[1], Öznur Pehlivan[2] and Ayten Yazgan-Karataş[2]*

[1]Molecular Biology and Genetics Department, Biruni University, Istanbul, Turkey
[2]Molecular Biology, Biotechnology and Genetics Research Center (MOBGAM) and Molecular Biology and Genetics Department, Istanbul Technical University, Istanbul, Turkey

*Corresponding author: karatasay@itu.edu.tr

DOI: https://doi.org/10.21775/9781912530328.01

Abstract

In nature, majority of bacteria live together in large complex sessile communities termed as biofilms. Within biofilm community, compared with their planktonic state, the physiological behavior of biofilm members is deeply altered, leading to gaining resistance to various environmental stresses, such as desiccation, antimicrobial agents and host immune response. As final stage of biofilm maturation, biofilm dispersal is also accepted as a critical step in bacterial lifestyle allowing dissemination of bacteria from the original site of infection in the host or the environment and increasing virulence of pathogenic bacteria. Biofilm formation and dispersal is under the control of complex regulatory mechanism that involves the coordination of various signaling messengers and molecular effectors. In this chapter, we review and summarize recent findings in the factors and regulatory pathways controlling the biofilm formation and dispersal in a range of Gram-positive bacteria, including some members of *Bacillus* and *Staphylococcus aureus*.

Introduction

Bacterial cells might undergo two different types of growth mode throughout their life cycles; one of which is their free-floating planktonic state while the second one is formation of sessile aggregates (Garrett et al., 2008). Although giving a final terminology to these aggregates did not come until late 70's, the great journey of understanding of the biofilms has started much earlier when Antoni van Leeuwenhoek observed "animalcules" on tooth surfaces (Bassler, 1999; Costerton, 1999). This discovery was the first evidence for microbial biofilm formation. It should be noted that biofilms are almost omnipresent in the nature and they are paid a good amount of attention since these structures can be found on surfaces that are at a contact with aqueous solutions, while they can be formed on the water-air barrier of aqueous environments as well (Hurlow et al., 2015; Vasudevan, 2014; Parsek and Singh, 2003). This specific characteristic allows them to be highly observed in industrial platforms, hospitals and on sanity-requiring places (Donlan and Costerton, 2002; Branda et al., 2001; Wijman et al., 2007). Biofilm composition, structure, formation, regulation mechanism and its dispersal has been widely studied and examined.

Biofilms are densely packed groups of microorganisms and all together these organisms produce extracellular polymeric substances (EPS) including proteins, enzymes, polysaccharides and DNA (Flemming and Wingender, 2010; Hall-Stoodley et al., 2004). Moreover, water is the main component of the biofilm and it enables the flow of nutrients in the biofilm matrix. Hence, the biofilm architecture comes from densely packed cells that have water channels for nutrient transport (Kostakioti, et al. 2013; Stanley and Lazazzera, 2004). Bacterial biofilms are normally beyond the reach of antibiotics and therefore not easily accessible by the human immune system. Even if these structures are reached by antibiotics, they have the potential to differentiate and nullify antimicrobial agents and therefore a long-term treatment becomes a requirement (Gilbert et al., 1997). Biofilm-forming bacteria can start a cascade of gene expression activation pattern which can eventually lead them to gain resistant phenotypes as a response to certain environmental changes such as cell density, availability of nutrients, temperature, pH and osmolarity (Fux et al., 2005).

Given that biofilm formation process happens through a cascade of events leading to adaptation under various nutritional and environmental conditions, it is possible to classify these events into a multi-step process (O'Toole and Kolter, 1998; Hall-Stoodley et al., 2004). As shown in Fig. 1, first event that signify the initiation of the biofilm formation would be the reversible attachment of planktonic cell to a surface by adhesion mechanism. This is followed by their irreversible association as microcolony on the surface, formation, by means of cell-cell signaling pathways, biofilm maturation and stabilization. Finally, some microorganisms detach from the biofilm and become planktonic cells until they colonize on another surface (Costerton et al., 1999).

EPS is predominantly composed of polysaccharides and other biomolecules such as proteins, lipids and nucleic acids (Costerton, 1999). Polymers such as glycopeptides, lipids and lipopolysaccharides form a scaffold and they coexist in the biofilm (Flemming and Wingender, 2010). Biofilm development and release of cells can be regulated by cell density-controlled gene expression. Global regulation system of cell density signaling in bacteria is also called as quorum sensing and those signaling molecules can be differentiated between Gram-positive and Gram-negative bacteria, such as acylated homoserine lactones (AHLs) for Gram-negative bacteria (Taga and Bassler, 2003) and small peptides for Gram-positive bacteria (Yarwood et al., 2004).

This chapter will shed a light on the environmental factors and regulatory mechanisms involved in the formation and dispersion in a range of Gram-positive bacteria that can produce biofilms, including some members of *Bacillus* and *Staphylococcus aureus*. The members of *Bacillus* genus that would be reviewed in this chapter will be primarily *B. subtilis* and two members of *Bacillus cereus sensu lato* group; *B. cereus sensu stricto* (will be metioned as *B. cereus*) *and Bacillus anthracis. Bacillus cereus sensu lato* represents a growing number of bacterial species. This group includes *B. anthracis* (anthrax agent), *B. cereus* (food poisoning), and *B. thuringiensis* (control insect pests) and along these mostly known species, there are also other less acknowledged representatives of this group including *B. mycoides*, *B. pseudomycoides*, *B. cytotoxicus*, *B. weihenstephanensis*, *B. wiedmannii*, *B. toyonensis*. Moreover, there are new additions to *Bacillus cereus sensu lato* group: *B. bingmayongensis*, *B. gaemokensis*, and *B. manliponensis* (Bazinet, 2017). But firstly, we will review the recent findings on the quorum sensing which is the central mechanism to regulate bacterial biofilm formation.

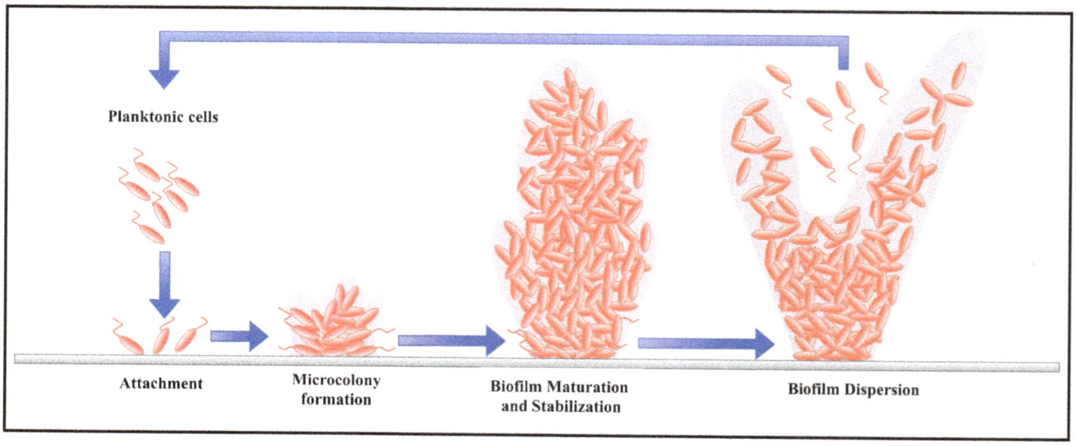

Figure 1. Life cycle of a biofilm. Biofilm formation is a sequential process requiring attachment of planktonic cells to a surface followed by microcolony formation and biofilm maturation with the emergence of the three-dimensional structures from which bacterial cells can disperse.

Cell to cell signaling in bacteria: Quorum Sensing

It is known that bacteria use multiple classes of chemical signals of different types to coordinate their behavior and to follow inter- and intraspecies communication. This hierarchical regulatory mechanism allows the bacterium to survive and interact with its natural habitat and this particular monitoring and regulatory system is called "Quorum Sensing" (QS) (Miller and Bassler, 2001; Daniels et al., 2004). In general, QS is based on the accumulation of signal molecules that increase in direct proportion to the population density and on the activation of the relevant receptor when the signal reaches a sufficient concentration.

The model Gram-positive bacterium *Bacillus subtilis* coordinates many adaptive behaviors such as chemotaxis, degradative enzyme secretion, antibiotic synthesis, competence development, biofilm formation and sporulation by using its QS ComQXPA signaling circuit which is used as a major quorum sensing pathway in *B. subtilis* (Auchtung et al., 2006; Comella and Grossman, 2005). This signaling circuit involves the interaction of four components: The pheromone processing enzyme ComQ, the peptide pheromone ComX, the histidine kinase receptor ComP and the response regulator ComA (Auchtung et al., 2006). *B. subtilis* uses the peptide ComX (5 to 10 amino acids) (Magnuson et al., 1994; Ansaldi et al., 2002), which is unique among bacterial peptide signals due to posttranslational isoprenylation of tryptophan. Isoprenylation on the tryptophan residue is essential for the active functioning of ComX (Ansaldi et al., 2002). Before processing and modification by ComQ isoprenyl transferase, ComX is synthesized at a length of 55 amino acids and following its conversion from a propeptide to a farnesylated active peptide, it is secreted into the extracellular medium (Bacon Schneider et al., 2002). When threshold concentrations are reached, it binds to the membrane-bound receptor kinase ComP and induces its autophosphorylation. Phosphorylated ComP (ComP~P) finally passes its phosphate group to the response regulator ComA, resulting in the transcription of the *srfA* operon that encodes the surfactin biosynthetic enzyme. Within this operon, the *comS* gene, which controls the competence development, is embedded. For this reason, both surfactin production and competence development are regulated together and strongly depend on the active ComQXPA QS system in *Bacillus subtilis* (Comella and Grossman, 2005).

Rap-Phr pairs also contribute to the *B. subtilis* QS system. In this system, Rap protein activity is coordinated with cell density via extracellular Phr signaling peptides. Phr peptides are synthesized as precursor proteins, Pro-Phr, containing an N-terminal signal sequence that is required for export into the extracellular environment by the SecA pathway (Pottathil and Lazazzera, 2003; Stephenson et al., 2003). Subsequently, extracellular proteases Subtilisin, Vpr and Epr process the Phr precursor into a mature penta or hexa-peptide (Lanigan-Gerdes et al., 2007). Mature Phr peptides are transferred into the cell by oligopeptide permease (Opp) (LeDeaux et al., 1997). Within the cell, Phr peptides bind to and inhibit the activity of their cognate Rap proteins (Perego, 2013). The *rap* and *phr* genes are generally genetically linked although not all *phr* genes have a cognate *rap* gene (Smits et al., 2007). All characterized Rap proteins contain two domains: a tricopeptide repeat (TPR) domain interacting with the signaling peptide and a second domain that functions either as a phosphatase or as a contact-dependent inhibitor of a response regulator (Perego, 2013; Gallego del sol and Marina, 2013). In domesticated strain *B. subtilis* 168, three response regulators that are targeted by the Rap proteins have been characterized as Spo0F, ComA and DegU (Omer Bendori et al., 2015).

In the Rap-Phr QS sytem, the activity of the major QS response regulator ComA is regulated by five chromosomally encoded Rap proteins, RapC, RapD, RapF, RapH and RapK. These proteins act as 'anti-activators' of ComA and they inhibit binding of this response regulator to DNA and eventually preventing the activation of target genes (Core and Perego, 2003; Bongiorni et al., 2005; Auchtung et al., 2006; Smits et al., 2007; Baker and Neiditch, 2011). Antagonistically, PhrC, PhrF and PhrK stimulate the activity of ComA by directly inhibiting the activity of their cognate Rap proteins RapC, RapF and RapK, respectively. All three peptides, PhrC, PhrF and PhrK are required for efficient expression of ComA-dependent genes. On the other hand, each stimulates the ComA-dependent gene expression at different extents (Auchtung et al., 2006). Among these peptides, PhrC (its mature-active form, also known as the competence and sporulation factor-CSF) has at least three different activities depending on its extracellular concentrations. At low concentrations (0.2-5 nM), CSF affects the expression of ComA~P-dependent genes positively by inhibiting the activity of RapC protein while its high concentrations (>20 nm) inhibits the phosphorylation of ComA through an unknown mechanism and thereby shifting the effect on the ComA dependent genes to a negative direction. Additionally, at high concentration, CSF stimulates sporulation by directly inhibiting another aspartyl phosphatase, RapB. Furthermore, RapB phosphatase dephosphorylates Spo0F~P, the intermediate response regulator in the *B. subtilis* phosphorelay pathway. This pathway is responsible for the phosphorylation and thus the activation of the transcription factor Spo0A, which is the essential regulator of both sporulation and matrix production in *B. subtilis* (Lazezzera et al., 1999; Branda et al., 2001; Hamon and Lazazzera, 2001).

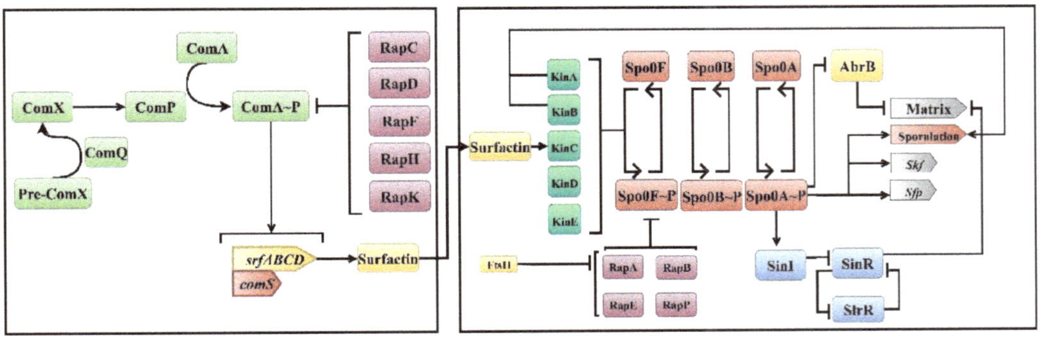

Figure 2. Schematic representation of QS ComQXPA regulatory circuit and its effect on biofilm development in *B. subtilis*. Proteins, open-reading frames (ORF), activation and repression patterns are indicated as rounded rectangles, triangles, arrows, and T-bars, respectively.

In *B. subtilis*, regulation of extracellular matrix (ECM) production (Lopez and Kolter, 2010) is combined with the QS system via two signaling molecules: ComX pheromone and lipopeptide antibiotic surfactin. Both molecules favorably enhance extracellular matrix production and thus cell differentiation in biofilm community (Lopez et al., 2009a; Lopez et al., 2009b) (Fig. 2). Surfactin secretion, as a surfactant, is required for swarming motility of *B. subtilis* cells (Kearns and Losick, 2003). Moreover, surfactin is also used as paracrine signal that trigger the creation of membrane pores and potassium leakage (Lopez et al., 2009b). This potassium leakage activates the phosphorelay kinase KinC, resulting in the phosphorylation of Spo0A (Spo0A~P) and thus activation of biofilm matrix genes. Actually, ComX and surfactin are implicated indirectly in the accumulation of Spo-0A~P (Lopez et al., 2009a). Increased Spo0A~P level causes the downregulation of pleiotropic regulator SinR, the major repressor of matrix genes: the *epsA-O* and *tapA-tasA-sipW* operons (Jiang et al., 2000; Kearns et al., 2005; Hamon and Lazazzera, 2001). This is achieved by Spo0A~P indirectly through the induction of the expression of SinI (Shafikhani et al. 2002) which then antagonizes the suppressor SinR (Bai et al., 1993). The expression of surfactin shows a bimodal pattern, that is, a few sets of cells in the population synthesize surfactin in response to ComX (Lopez and Kolter, 2010; Lopez et al., 2009b). In this circuit, ComX triggers the surfactin production. In turn, surfactin leads a subpopulation of cells to produce an extracellular matrix. Cells that produce surfactin cannot respond to surfactin and once surfactin-responsive cells initiate matrix production, they no longer respond to ComX by means of the presence of extracellular matrix itself and thus cannot become surfactin producers themselves. Moreover, in *B. subtilis*, the entry to sporulation can be delayed by inducing cannibalism through toxin production (Gonzalez-Pastor et al., 2003). Cannibals produce Skf and Sdp toxins and low levels of Spo0A~P induce the expression from cannibalism operons *skfABCEFGH* and *sdpABC*, so that sporulation can be delayed subsequently (Chen et al., 2006; Fujita et al., 2005). Therefore, nutrients released by cannibalized cells are likely to be used by matrix-producing cells and sporulation would be delayed eventually (Lopez et al., 2009c).

In *B. subtilis*, bacterial cell-cell communication is also shown to be mediated by LuxS/AI-2-dependent QS. *B. subtilis* follows a LuxS-dependent QS system as well, that regulates its morphogenesis and social behavior (Lombardia et al., 2006). *B. subtilis luxS* is a growth phase-regulated gene whose expression is under the control of AI-2 negative regulatory feedback loop. Although LuxS do not affect spore development, AI-2 production is regulated negatively by the main regulatory proteins, SinR and Spo0A. Interestingly, wild *B. subtilis* cells are proposed to require LuxS-dependent QS to form stable and differentiated biofilms on solid surfaces. LuxS activity is also required for the formation of sophisticated air colonies that act as big fruiting bodies in which AI-2 production and spore morphogenesis are regulated on a location-specific manner inside the developing colony. Hence, it is suggested that LuxS/AI-2 might create a new form of quorum sensing regulation in which AI-2 acts as a morphogen-like molecule that coordinates the social and multicellular behaviour of *B. subtilis* (Lombardia et al., 2006; Hardie and Heurlier, 2008).

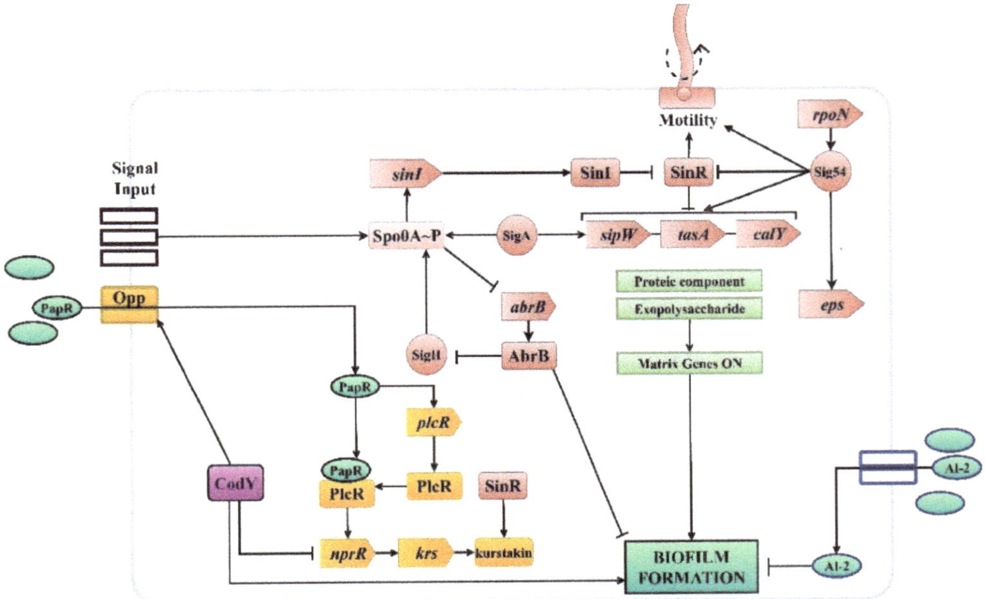

Figure 3. Schematic overview of the complex regulatory network which controls biofilm formation in *B. cereus*. Proteins, open-reading frames (ORF), signal peptides, sigma factors, activation and repression patterns are indicated as rounded rectangles, triangles, ellipses, circles, arrows, and T-bars, respectively. Flagella is represented as red and curvature represents rotation of the flagella.

Communication systems in other *Bacillus* spp. are also based on the secretion and recognition of cell-cell signaling molecules. In *Bacillus cereus*, the PlcR/ PapR quorum sensing system controls the expression of genes encoding virulence factors including degradative enzymes (Agaisse et al., 1999; Slamti and Lereclus, 2005) (Fig. 3). PlcR is activated by binding to a signal peptide, PapR, produced as a propeptide under the control of PlcR and this system is basically alike with ComQ-X system found in *B. subtilis*. Similarly, PapR becomes a target for the extracellular processing and it becomes processed into an active heptapeptide (Bouillaut et al., 2008). Thereon, it is transferred back into the bacterial cell through OppABCDF, oligopeptide permeation system (Gominet et al., 2001). Within the cell, PapR interacts with PlcR, a transcriptional regulatory protein and the resulting complex activate the PlcR regulator (Gohar et al., 2008). This regulatory protein of 48 amino acids consists of two separate domains; a DNA binding domain and a peptide binding domain. Peptide-binding-domain of this protein consists of five tetratricopeptide repeats (TPR). TPR domain interacts with a secreted signal peptide and this interaction leads to the activation of a new family of central regulatory competence sensor (RNPP family) which is only found in Gram-positive bacteria with low G + C content (Declerck et al., 2007). These sensors include NprR from *B. cereus* (Perchat et al., 2011) and RAP phosphatases from *B. subtilis* (Parashar et al., 2011; Ivanova et al., 2003).

PlcRa, a PlcR paralogue, is also a part of QS system in *B. cereus* (Huillet et al., 2012). PlcRa is suggested to play a role during the oxidative stress response and cysteine metabolism during the early stationary phase. In particular, PlcRa activates the expression of various genes involved in the pathway of cysteine synthesis from methionine, including the S-adenosylmethionine (SAM) recycling pathway. This transcriptional control of the SAM recycling pathway by PlcRa affects another QS system, AI-2-dependent system. The role of AI-2 in *Bacillus subtilis* QS system is already discussed and it is known that this signaling metabolic molecule is produced by LuxS, which is also a crucial metabolic enzyme involved in the recycling of SAM that is found in many bacterial species and is therefore proposed to be important for communication between species (Schauder et al., 2001). *B. cereus* synthesizes and recognizes AI-2 as an extracellular signal and inhibits biofilm formation when a concentration threshold of this molecule is reached. CymR regulon in the global regulation of gene expression in response to the presence of cysteine is also suggested to be essential for AI-2 production in combination with the PlcRa system in *B. cereus* (Huillet et al., 2012).

The relatively sensitive arrangement of genes, which are controlled over a density-dependent manner, has been described for many species of bacteria including *Bacillus anthracis* (Jones et al., 2005). In accordance with *B. cereus* and *B. subtilis*, there is a functional *luxS* ortholog in *B. anthracis*, which is required for the growth phase specific AI-2 expression (Jones and Blaser, 2003).

Two-component regulatory systems are not only common in *Bacillus spp.*, they are also used by other Gram-positive species like *Staphylococcus aureus* to adapt to their environments. In common with *Bacillus spp.* systems, *S. aureus* two-component regulatory system consists of two proteins: a histidine kinase and a DNA binding response regulator. When a signal is received and sensed by the histidine kinase component, it gets autophosphorylated at that point and then passes its phosphate group to its cognate response regulator (Yarwood et al., 2004). Ultimate cellular response is then derived by phosphorylated response regulator, since it also acts as a transcriptional regulator. Although the mechanisms are similar for different Gram-positive organisms, components that play a role are different. In *S. aureus*, SaeRS takes an important part in controlling the production of more than 20 virulence factors including hemolysins, surface proteins, and proteases. SaeS is the before-mentioned histidine kinase of the system, whilst SaeR is the response regulator. This two-component regulatory system receives and creates a response against both intra- and inter-species signals (Novick and Jiang, 2003; Geiger et al., 2008; Makgotlho et al., 2013). SaeRS is encoded from the *saePQRS* operon and SaeP and SaeQ are the two auxiliary proteins. From the *saePQRS* operon, which has an autoregulated promoter, SaeR and SaeS are encoded together (Novick and Jiang, 2003; Jeong et al., 2011). SaeR is a member of OmpR family response regulators and once phosphorylated, it can bind DNA. Despite this binding ability, it still shows different affinities against varied target promoter regions. Hence, the SaeRS regulon is divided into two different classes. First class contains the genes that are targets of this regulon at a lower affinity, therefore a high concentration of phosphorylated SaeR is needed to derive an action from these genes. Class II genes are high affinity targets, such that even at lower concentrations, phosphorylated SaeR can exert a regulatory activity on their promoters (Mainiero et al., 2010). Although SaeRS is a very important two-component response regulator system in *S. aureus*, it is not the only one, per se. The QS in *S. aureus* is also regulated by the *agr* operon. *agrD* from this operon is responsible for encoding an autoinducible peptide (Yarwood et al., 2004) (Fig. 4). Sensor histidine kinase AgrC follows the conventional QS signaling paths of Gram-positive bacteria and its cognate response regulator, AgrA. The membrane-bound protein AgrB is not only responsible for the modification of prepeptide AgrD but also for its export. Once a threshold concentration of AgrD is attained extracellularly, the mature peptide AgrD is sensed by AgrC. It should be noted, there is not a permease involved in the import of AgrD, which is in contrast with the other QS regulation systems. Since the membrane is not permeable to AgrD, the response system is activated by

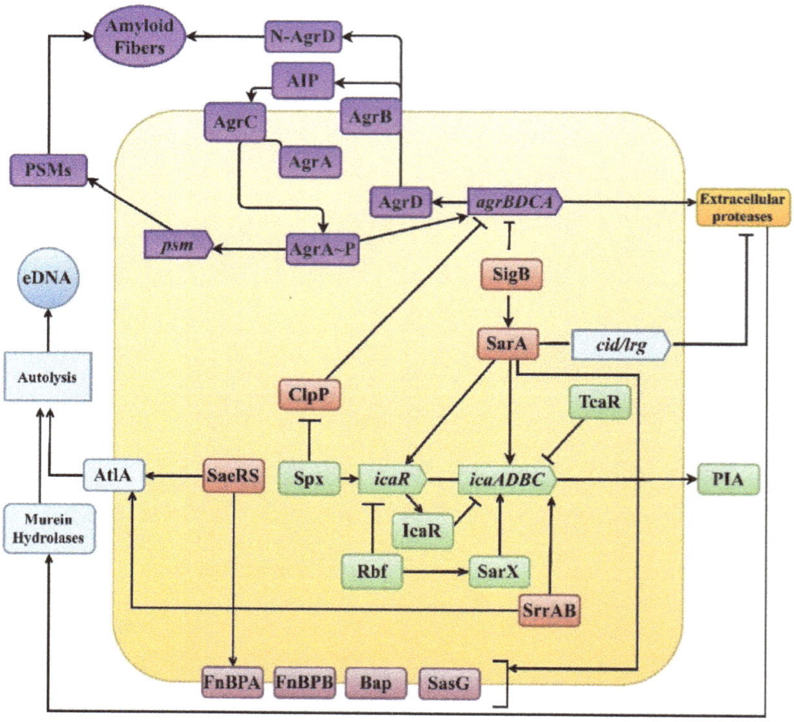

Figure 4. Schematic overview of the regulatory network which controls biofilm formation in *S. aureus*. Proteins and peptides, open-reading frames (ORF), eDNA, amyloid fibers, activation and repression patterns are indicated as rounded rectangles, triangles, circle, ellipse, arrows, and T-bars, respectively. Autolysis mechanism is represented as rectangle.

the autophosphorylation of the membrane bound AgrC upon sensing AgrD concentration that is accumulated outside. Correspondingly, the phosphorylation cascade keeps rolling and AgrA response regulator gets activated in turn. Phosphorylated AgrA activates its own transcription and the transcription of other targets such as regulatory RNA, RNAII and effector RNA, RNAIII (Novick and Geisinger, 2008; Chan et al., 2004). *S. aureus* QS regulation of biofilm development is assumed to depend on the Agr system (Cheung et al., 2004) and activation for RNAIII expression leads to biofilm maturation and dispersal (Karatan and Watnick, 2009).

Regulation of Biofilm Development

B. subtilis is considered to be an important model organism in terms of cellular differentiation since it has the ability to develop motility, form biofilm, develop competence and sporulate. When studied under laboratory conditions, *B. subtilis* biofilm formation is observed to exist at different areas, including air-solid interface on agar plates, called as colony biofilm, air-liquid interface where they hang without an attachment, termed as floating biofilm/pellicle. Finally, they can also survive and form colonies at liquid-solid interface where they adhere to the edges of the vessels they are submerged in (Branda et al., 2001; Kobayashi, 2007a). On the air-solid interface such as agar surface, biofilms appear as complex wrinkled colonies (Hamon and Lazazzera, 2001). Wrinkled structure can emerge because of the localized cell deaths and the rigidness of ECM which starts to fold on itself by the shear mechanical force (Asally et al., 2012; Trejo et al., 2013). Wrinkling presents the bacteria some advantages. These advantages include the increase of the surface-to-volume ratio and formation of liquid channels, so that bacteria can quench the need for oxygen, since channels can make the penetration of liquids, gas and nutrient easier (Dietrich et al., 2008; Kolodkin-Gal et al., 2013; Wilking et al., 2013). Still, biofilm wouldn't be able to stop the natural ageing process and sporulated cells would start to become apparent on the surface of the biofilm (Branda et al., 2001; Hamon and Lazazzera, 2001). It should be noted that although motility of the cells gets inhibited when cells grow more stacked in biofilms, still a few of the cells keep their motility without losing their flagella (Vlamakis et al., 2008).

Furthermore, with the help of these motile cells and release of surfactant molecules, biofilms keep expanding (Angelini et al., 2009). Hence, within biofilms, a series of morphological changes occur, and several different cell types can coexist in genetically similar *B. subtilis* populations. These differentiated cells might be composed of subpopulations of motile cells, matrix-producing cells and dormant spores (Lopez and Kolter, 2010). In *B. subtilis,* the transition from the planktonic cells into sessile cells is accompanied by morphological changes (Branda et al., 2001; Kobayashi, 2007a). In the early steps of biofilm formation, most cells are in planktonic form and flagellar genes, their expressions are σ^D- dependent, are actively expressed in *B. subtilis*. Furthermore, biofilm formation begins with the formation of cell chains that are constructed by suppression of *B. subtilis* alternative sigma factor, σ^D and σ^D-dependent autolysis (Chen et al., 2009). On the other hand, functional regulators, σ^H, Slr, and Spo0A are also required for the formation of these cell chains (Lemon et al., 2008; Branda et al., 2006; Kobayashi, 2007a). Additionally, regulators-Abh, AlsR, σ^X, YvrH, and CysL are essential for cell chains to form bundles. As the number of cell chains increases, the chains form clusters. The formation of cell clusters depends on the production of extracellular matrix (Kobayashi, 2007a).

In *B. subtilis*, these morphologically differentiated cell types are mediated and regulated by precise levels of certain sets of global regulators, Spo0A, SinR and DegU (Hamon and Lazazzera, 2001; Kearns et al., 2005; Chu et al., 2006; Kobayashi, 2007b; Verhamme et al., 2007) (Fig. 5). Cell heterogeneity and cell differentiation in biofilms are partly regulated by Spo0A, the master transcriptional regulator in *B. subtilis* (Fujita et al., 2005; Molle et al., 2003). Spo0A regulates the expression of a large number of genes depending on the concentration of its active form, phosphorylated Spo0A (Spo0A~P) (Fujita et al., 2005). Low levels of Spo0A~P lead to the induction of the matrix genes expression (Kearns et al., 2005), whereas, high levels of Spo0A~P represses matrix genes and induces sporulation genes. Thus, when Spo0A gets phosphorylated, biofilm formation becomes induced. As the biofilm matures, Spo0A~P starts to accumulate in matrix-producing cells and they proceed with the sporulation (Fujita et al., 2005; Vlamakis et al., 2013).

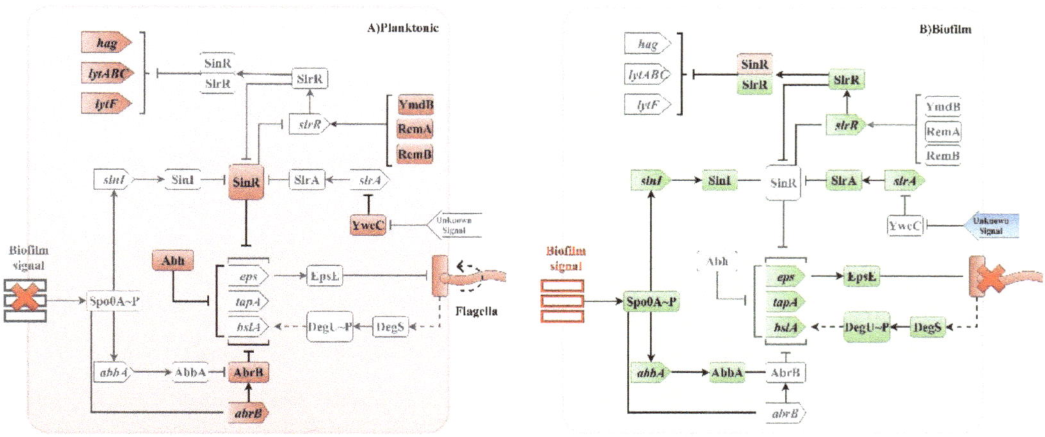

Figure 5. Schematic overview of the regulatory circuits which control gene expression during (A) planktonic growth and (B) growth as a biofilm. Proteins, open-reading frames (ORF), activation, indirect activation and repression patterns are indicated as rounded rectangles, triangles, arrows, dashed arrows, and T-bars, respectively. Red represents active genes and proteins in planktonic cells while green represents active genes and proteins in biofilm. Flagella is represented as red, curved dashed arrow and cross represent rotation of the flagella and inhibition of flagellar rotation. All inactive genes, proteins and interactions are indicated as gray.

The activity of Spo0A is regulated through phosphorylation which involves at least 4 kinases; KinA, KinB, KinC and KinD. Nonetheless, this phosphorylation process cannot be determined directly by these kinases, even though, they can exert their effect indirectly through a phosphorelay mechanism. It is thought that the stepwise activation of Spo0A~P through phosphorelay is provided by the differential activation of these histidine kinases in *B. subtilis*. The phosphorelay starts with the activation of Spo0F when it gets phosphorylated by either KinA-D. Conveniently, throughout the phosphorelay mechanism, the phosphate group is transferred from Spo0F, to Spo0B and from Spo0B finally to Spo0A. There is also another kinase, KinE, which is proposed to contribute to phosphorelay mechanism. However, its role in the matrix gene expression is unclear (McLoon et al., 2011). KinC and KinD proteins have also been shown to induce biofilm development under aerobic conditions (Mhatre et al., 2014), while the ability to move is dependent on KinB and KinC. Besides, KinA and KinB are responsible for providing the main input for sporulation initiation but controversially they still contribute to biofilm formation (Grau et al., 2015) (Fig. 2).

Surfactin secretion is required for swarming motility of *B. subtilis* cells (Kearns and Losick, 2003). Nonetheless, it is also used as a paracrine signal that triggers the microdomains (a feature like lipid rafts) and potassium leakage (Lopez et al., 2009b). This potassium leakage can be created not only by surfactin but also by fungicide nystatin and antibiotic valinomycin (expressed by co-resident species). This phenomenon activates the phosphorelay kinase KinC, resulting in the phosphorylation of Spo0A and thus activation of biofilm matrix genes. Moreover, chlorine dioxide is shown to activate matrix gene expression via KinC, this time not by creating microdomains but by decreasing membrane potential (Shemesh et al., 2010). On the other hand, KinD, described as a "checkpoint" protein, modulating the conversion of matrix producers to spore formers, displays dual function: it exerts its phosphatase activity until biofilm matrix is sufficiently produced and then displays its kinase activity that contributes to attainment of high level of Spo0A~P, which is required for sporulation (Aguilar et al., 2010).

Major component of the phosphorelay system, Spo0F can be repressed by RapA, RapB and RapE (Veening et al., 2005). In addition, it was also found that the plasmid pBS32 contained in *B. subtilis* NCI3610 strains carried an additional *rap-phr* cassette called *rapP-phrP* which is required for the biofilm formation (McLoon et al., 2011). Subsequently, in addition to regulating *B. subtilis* biofilm formation, it was shown that RapP regulates sporulation and genetic competence as a consequence of its ability to dephosphorylate Spo0F. RapP-mediated dephosphorylation of Spo0F would drain the phosphorylated level of Spo0A. This situation would result in the decreased expression of Spo0A-dependent sporulation gene *spoIIG* and increased expression of the *epsA-O* operon, responsible from EPS production, and *srfA* operon including the genes important for biofilm formation and competence development (Parashar et al., 2013). Furthermore, membrane microdomains are also proposed to be essential to activate biofilm formation in *B. subtilis* and robustness of these microdomains are attained via stabilization of KinC and the membrane-bound protease FtsH that cleaves specific Rap phosphatases (Yepes et al., 2012; Mielich-Suss et al., 2013).

The regulatory network that controls cell differentiation is not only limited to Spo0A action. DegU (Kobayashi, 2007b) and ComA (Tran et al., 2000) are also key regulatory elements that leads the cells to acquire different lifestyles inside the biofilm. DegU plays an important role in the formation of biofilm in *B. subtilis* (Fig. 6). The gradual phosphorylation of DegU determines the faith of the cells. In its unphosphorylated state, DegU activates the expression of ComK, which is the global regulator of competence development in *B. subtilis*. Once phosphorylated, DegU~P activates motility genes (low level of DegU~P), genes related with biofilm formation (intermediate level of DegU~P), or exoprotease production (high level of DegU~P) (Kobayashi, 2007b; Verhamme et al., 2007). Moreover, DegU~P enhances the expression of BslA protein, a key component of extracellular matrix of biofilm in *B. subtilis* (Kobayashi and Iwano, 2012; Hobley et al., 2013). Dephosphorylation of DegU can be directed by ClpC-mediated proteolysis (Ogura and Tsukahara, 2010). Furthermore, RapGH is responsible for the dephosphorylation of DegU~P. Phosphorylation event can be controlled by DegQ, which stimulates phosphotransfer from DegS to DegU (Veening et al., 2005). DegQ expression is regulated by ComQXPA, QS system. Thus, QS signals can stimulate the growth mode transition by phosphorylating DegU.

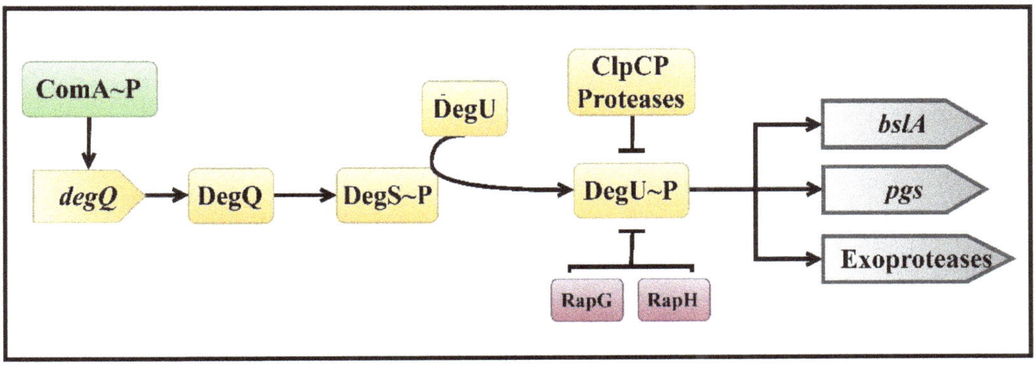

Figure 6. Schematic representation of DegU regulatory circuit which modulates biofilm formation in B. subtilis. Proteins, open-reading frames (ORF), activation, and repression patterns are indicated as rounded rectangles, triangles, arrows, and T-bars, respectively.

As mentioned above, biofilms are formed by population of microorganisms that live together in the extracellular matrix that they produce. In *B. subtilis*, this extracellular matrix (ECM) is composed of the exopolysaccharide (EPS) encoded by the 15 genes of *epsA-O* operon (Branda et al., 2001; Branda et al., 2006; Romero et al., 2010). Among the gene products of the *eps* operon, EpsE has bifunctional role that coordinates the production of EPS when motility of the cells decreases (Guttenplan et al., 2010). EpsE behaves as a clamp that inhibits the flagellar movement via interacting with FliG, flagellar motor protein. It also exhibits glycosyltransferase activity involved in the synthesis of EPS. Due to this mechanism, cells give up the motility when matrix production is initiated. Not only mutations in the *eps* genes, but also in *pgcA* and *gtaB* genes result in deteriorated biofilm formation, since these latter two encode α-phosphoglucomutase and UTP-glucose-1-phosphate uridylyltransferase, respectively. These two enzymes participate to the synthesis of nucleotide sugars, which are precursors of the *eps* pathway (Lazarevic et al., 2005; Branda et al., 2004).

In *B. subtilis*, two structural proteins TasA and BslA have been well identified in the biofilm matrix. TasA (formerly YqxM) is assembled into long amyloid fibers and anchored to the cell wall by the TapA protein which is found in the cell wall fraction of biofilm cells (Romero et al., 2010; Romero et al., 2011). In addition, TapA can be recovered at a very low amount from the amyloid fibers (Romero et al., 2011). The type I signal peptidase SipW, which is also encoded by the *tapA-sipW-tasA* operon, is required for processing and transport of TasA and TapA to the extracellular biofilm matrix (Stover and Driks, 1999; Tjalsma et al., 2000). SipW acts on TapA and TasA by recognizing their N-terminal signal sequence which leads to the cleavage and release of these proteins from the membrane and ultimately they are converted into cell-wall associated fibers. In addition to the processing of TasA and TapA protein, SipW also plays a regulatory role which is specific to the surface-adhered biofilms (Terra et al., 2012). It was shown that SipW is a bifunctional protein, apart from its signal peptidase activity, SipW activates the biofilm matrix genes (the *eps* operon) through its C-terminal domain. But SipW dependent-activation of *eps* genes occurs only when cells are present on a solid surface. The *B. subtilis* biofilm matrix also contains the structural protein, BslA (Biofilm surface layer) which forms a hydrophobic protective layer surrounding the biofilm (Vlamakis et al., 2013; Hobley et al., 2013). As mentioned above, the expression of *bslA* gene is activated by DegU~P. Moreover, *bslA* expression is also upregulated by LutR regulator protein via direct binding (Irigul-Sonmez et al., 2014). Strains that are mutated on *lutR* gene are also showing a distinct phenotype since *bslA* and *tapA* expressions are shown to be downregulated by the lack of LutR. Hence, it is proposed that LutR is also significant in biofilm formation in *B. subtilis* (Irigul-Sonmez et al., 2014). ECM not only contains proteins like TasA and BslA but it also consists of another extracellular polymer, γ-poly-DL-glutamic acid (PGA) (encoded from *pgs* operon) which is required for submerged biofilm formation in *B. subtilis* (Kobayashi, 2007a; Branda et al., 2006; Stanley and Lazazzera, 2005; Morikawa et al., 2006). Notably, beside to BslA, PGA production is also regulated by DegU~P in *B. subtilis* (Stanley and Lazazzera, 2005; Morikawa et al., 2006).

Another crucial protein for the biofilm development is the SinR protein which regulates the transcription of the *epsA-O* and *tapA-sipW-tasA* operons and activates the transcription of flagellar genes (Kearns et al., 2005; Chu et al., 2006). SlrR, another protein that is negatively regulated by SinR, induces transcription of the *tapA-sipW-tasA* operon (Chu et al., 2008). Thus, the SinR protein promotes cell motility and differentiation and provides the ability to switch back and forth between the planktonic and biofilm forms of the *B. subtilis* lifestyle (Kearns et al., 2005) (Fig. 5). As mentioned above, activity of SinR is regulated by SinI during stimulation of biofilm formation and Spo0A induces the expression of SinI, an antagonist repressor of SinR (Shafikhani et al., 2002). The interaction between SinR and SinI is also modulated by YmcA and YlbF proteins (Kearns et al., 2005). It is known that the regulation of SinI, YlbF and YmcA activity contributes to the transition from planktonic to biofilm-forming growth mode by controlling the activity of SinR (Kearns et al., 2005; Chu et al., 2006; Kolter and Greenberg, 2006). In *B. subtilis*, the expression of matrix components, extracellular polysaccharide and secreted protein TasA is suppressed by the DNA binding master regulator SinR. Along with SinR, two small protein regulators of the extracellular matrix are also identified: RemA (YlzA) and RemB (YaaB). The RemA or RemB deficiency is shown to disrupt the pellicle formation and complex colony formation. Both proteins are required for the activation of matrix biosynthetic operons and for the inhibition of motility genes. Besides, they behave in parallel with SinR and the other two biofilm regulators, AbrB and DegU (Winkelman et al., 2009).

It is already mentioned that the response regulator Spo0A controls many events that occur in stationary phase in *B. subtilis*. Low level of phosphorylation on Spo0A also suppress the transcription of *abrB*, which encodes a transcriptional repressor. It is well established that the major role of Spo0A~P in biofilm formation is also provided by the suppression of *abrB*, since it is the downregulator of the epsA-O and the *tapA-sipW-tasA* operons required for biofilm formation (Hamon et al., 2004). AbrB acts synergistically with AbrB-homolog repressor protein Abh and an AbrB inhibitory protein, AbbA and all together they are responsible for mediating the fine tuning of matrix-related gene expression (Strauch et al., 2007; Banse et al., 2008). Furthermore, AbrB represses expression of the matrix protein BslA (Verhamme et al., 2009) and together with AbrB, SinR represses the regulatory proteins SlrR (Chu et al., 2008).

The mutation in the flagella genes has been shown to cause delayed pellicle formation in a rich medium. (Kobayashi, 2007a). This phenotype suggests that flagella plays a regulatory role in biofilm formation. Mutations in late-flagella genes have also been shown to reduce SlrR activity, which is required for the growth-mode switch (Fig. 5). SlrR is a member of the LexA family of autopeptidases, and it is crucial for biofilm formation since it can bind to SinR, holding it back from repressing *epsA-O, tapA-sipW-tasA* and *slrR* itself. Therefore, SinR-SlrR complex creates a regulatory loop such that SinR concentration would be high when SlrR concentration is low and vice versa. Furthermore, the SinR-SlrR complex is also responsible for the repression of *hag* gene (encodes flagellin) and *lytABC* and *lytF* genes (encoding autolysins) (Chai et al., 2010a; Chai et al., 2010b). SlrR is not a stable protein and therefore, it can be subjected to self-degradation or it can be cleaved by ClpCP protease. Accordingly, this instability results in the decrease of SlrR concentration eventually and hence removes the repression on autolysin encoding genes. This leads to delayed sporulation on the cells that are parts of colonies and pellicles (Chai et al., 2010a) (Fig. 5).

The question of "When the cells will have low SlrR or low SinR level?" is responded via SinI since it competes to bind to SinR, just like SlrR. Additionally, SlrR is an antagonist of SinR, so that it binds to and inhibits the SinR repressor. Thus, the production of SinI inhibits SinR, leading to the depression of the gene for SlrR. This results in the accumulation of SlrR to high levels and further inhibition of SinR by SlrR, driving the switch into SlrR high state. This epigenetic control is also regulated by YwcC and SlrA. SlrA functions similarly to SinI (Chai et al., 2010a) and it is repressed by the transcription repressor YwcC. Expression of SlrR is also enhanced by Abh, which is repressed by AbrB (Strauch et al., 2007). Along with Abh, YmdB (a putative phosphoesterase) also enhances SlrR production. In addition, *slrR* gene expression also requires RemA and RemB both of which act in conjunction with SinR (Winkelman et al., 2009).

In *B. subtilis*, AI-2 production is also regulated by Spo0A and SinR and spatial activity of LuxS is important for the development of specific biofilm assemblies (Lombardia et al., 2006). Moreover, next to the above mentioned transcription factors, *epsA-O* operon is also regulated by a cis-acting RNA element EAR (*eps*-associated RNA element) which is thought to function as an anti-terminator since it is embedded in *eps* operon and it is proposed that it might enhance expression of *eps* genes by interacting with RNA polymerase (Irnov and Winkler, 2010).

Although these regulator systems Spo0A/AbrB, SinR/SinI, SlrR/SlrA and DegS/DegU are known to react to different environmental signals, the specific signals that each system responds to are still unclear. Another question that remains is whether these regulators work independently and whether there will be another regulatory system to synergistically cross growth models or to coordinate their actions. Analysis of the interactions between these regulators is also necessary to understand the mechanism by which biofilm formation is initiated.

Up until now, a very complex and intertwined multifunctional regulatory network of many gene products is reviewed. On the other hand, it should be noted that many extracellular triggers are playing important roles in biofilm development in *B. subtilis*. Recent studies show that magnesium ions can reduce the biofilm formation of *Bacillus* species by downregulating the expression of matrix genes. *epsA-O* and *tapA* operons, as well as Spo0A (Oknin et al., 2015). Along with magnesium, calcium is also presented to reduce expansion of *B. subtilis* colony biofilms (Mhatre et al., 2017). These cells in the biofilm colony secrete extracellular polysaccharides (EPS) and protein components, which hold them together and provide a protective hydrophobic shield. The cells also release surfactin that help to colonize solid surfaces and Ca^{+2} is suggested to modify the amphiphilic properties of surfactin and inhibit biofilm colony expansion (Mhatre et al., 2017).

Furthermore, *B. subtilis* can use a wide variety of sugars, organic acids and other organic compounds as carbon and energy source. Traditionally, glucose is the preferred choice, although alternative carbon sources can be metabolized under glucose depleting conditions (Fujita, 2009). Previous studies have shown that catabolite expression and activation in *B. subtilis* are both dependent on the transcription regulator, catabolite control protein A (CcpA) and HPr and Crh components, all of which are controlled by HprK kinase/phosphorylase (Lorca et al., 2005). It has been recently discovered that malate serves as a second preferred carbon source for *B. subtilis* (Meyer et al., 2011) and extracellular glucose and catabolite repression exert different effects on biofilm formation in *B. subtilis*. Biofilm formation is activated when glucose is found at low concentrations but inhibited when glucose is present at high concentrations (Stanley et al., 2003).

Along with glucose, galactose metabolism plays an important role in biofilm formation in *B. subtilis*. The three main enzymes that make up Leloir pathway are GalK, GalT and GalE and galactose metabolism is dependent on this pathway. It is reported that in the absence of GalE, *B. subtilis* develops sensitivity to exogenous galactose, which leads to rapid cell lysis. Galactose metabolism genes play an important role in the formation of *B. subtilis* biofilm, and that the expressions of both *gal* and *eps* genes are interrelated (Chai et al., 2012). In addition, butyric acid (BA), which is released in the lipolysis of milk fat is a powerful trigger for biofilm bundle formation. BA-induced biofilm formation is mediated by histidine kinase KinD (Pasvolsky et al., 2014). Strikingly, c-di-GMP-signaling plays a role in the control of swarming motility but not in biofilm formation in *Bacillus subtilis* (Chen et al., 2012; Gao et al., 2013).

Although *B. subtilis* is the most broadly studied species of *Bacillus spp.* for biofilm formation, it sure is not the only one. Phylogenetic analysis of the *Bacillus* genus shows that *B. cereus* is the most closely related subgroup to *B. subtilis* and both of these species are known to produce biofilms, for *B. cereus,* preferentially at air-liquid interfaces (Wijman et al., 2007). Even though they are closely related and these two biofilms forming organisms share some of the molecular mechanisms involved in the formation and regulation of biofilm, there are sharp differences between them with regard to structural variations, matrix formation and regulatory pathways.

Flagella development and hence mobility in *B. cereus* is important for the formation of biofilms to reach sites and to adhere to epithelial cells (Houry et al., 2010; Ramarao and Lereclus, 2005). It is known that chemical composition of the cell surface affects biofilm formation and maturation (Candela et al., 2011). Reversible attachment of *B. cereus* to the surface is often directed by physicochemical interactions and is influenced by cell surface properties such as hydrophobicity and electrical charge (Karunakaran and Biggs, 2011). Molecules capable of modifying cell surface properties, such as hypothetical cell wall peptidase (CwpFM) also affect biofilm formation in *B. cereus*. Those peptidases are proposed to function in peptidoglycan modification, motility and adhesion (Tran et al., 2010). In addition, cell surface properties are altered during biofilm formation. For example, *B. cereus* biofilm cells are more hydrophilic, they display reduced surface charge and altered composition of cell wall polysaccharides (Karunakaran and Biggs, 2011). Other components of biofilm include extracellular polymeric substances (EPS) which contain proteins, polysaccharides, lipids, and extracellular DNA, but the composition and relative amounts may vary depending on the species and strains (Abee et al., 2011).

In *B. subtilis*, biofilm formation is controlled by several transcriptional regulators mentioned previously. Some of these regulators are conserved in the *B. cereus*, including the major transcriptional repressor SinR (Kearns et al., 2005; Pflughoeft et al., 2011; Fagerlund et al., 2014). SinR is antagonized by SinI in both species. In case of SinI deficiency, biofilm and hypermotility are observed while in case of *sinR* deletion, biofilm overproduction and a decrease in motility occurs. There is no paralog of *bslA* or *tapA* in the *B. cereus* genome, but similarly, in pathogenic *Bacillus cereus* or *Bacillus anthracis* strains, *tasA* paralog is included in the *sipW-tasA* operon (Caro-Astorga et al., 2015) (Fig. 7). CalY1 and CalY2 are members of a metalloproteinase family proteins, usually termed camelysins and *calY1* and *calY2* genes are considered as *tasA* like genes and moreover, they are located next to *sipW-tasA* (Gao et al., 2015; Caro-Astorga *et al*., 2015).

Similar to *B. subtilis*, Spo0A also plays a key regulatory role for biofilm formation in *B. cereus* (Gao et al., 2015). Along with *spo0A-sinI-sinR* regulatory agents, which display central roles in biofilm formation, cell differentiation, and bacteria-host interactions in *B. cereus*, other regulator proteins that affect *B. cereus* biofilm formation are studied although the exact regulatory pathways of some of them are still unknown (Majed et al., 2016) (Fig. 3). These include biofilm inhibitor AbrB (Fagerlund et al., 2014), SigH, virulence regulator PlcR (Hsueh et al., 2006) and necrotrophic regulator NprR (Dubois et al., 2012). CodY is another response regulator whose homolog is also present in *B. subtilis* and whose roles are sensing the nutrient availability and conducting stationary phase adaptations (Hsueh et al., 2008, Frenzel et al., 2012). Among the responsibilities that CodY carries is the activation of virulence factors and PlcR (Frenzel et al., 2012; Lindback et al., 2012). In turn, PlcR activates NprR at the onset of the stationary phase (Dubois et al., 2013), promoting the production of kurstatin in *B. cereus* and being a biosurfactant lipopeptide, kurstakin promotes *B. cereus* biofilm formation (Gelis-Jeanvoine et al., 2017; Dubois et al., 2012).

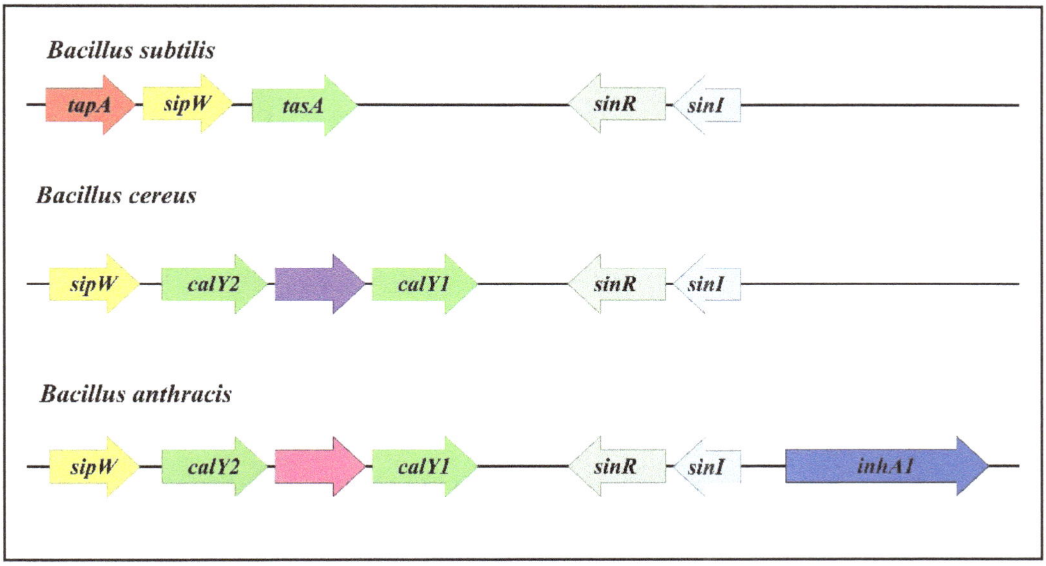

Figure 7. Schematic representation of the *sin* loci of *B. subtilis*, *B. cereus* and *B. anthracis*: Open reading frames are represented as block arrows and ortholog genes are highlighted in same color.

Furthermore, as a difference, it is also shown that the *eps* locus, which is homologous to *epsA-O* operon of *B. subtilis* is not effective for pellicle formation in *B. cereus*, while the deletion of *epsA-O* results in fragile biofilm pellicle formation in *B. subtilis* (Lemon et al., 2008). Interestingly, anti-termination RNA element named EAR is observed both in *epsA-O* of *B. subtilis* and in the *eps* locus of the *cereus* group, rather signaling an expected similarity (Irnov and Winkler, 2010). Still, although *B. subtilis epsA-O* operon is required unconditionally for biofilm formation, deletion of the *B. cereus eps* locus does not affect biofilm formation (Gao et al., 2015). Hence, the origin of polysaccharides in the *B. cereus* biofilm matrix is still unknown. Furthermore, DegU global response regulator, has no paralog in *B. cereus* while the *plcR* virulence regulator mentioned above, which is responsible for regulating the expression of many genes in *B. cereus*, has no paralog in *B. subtilis* (Gohar et al., 2008).

In addition to the regulatory effects of the transcription factors, biofilm formation in *B. cereus* is also induced by the signaling molecule cyclic-di-GMP (Fagerlund et al., 2016). Transcription regulator Sigma 54 (RpoN) is another regulatory protein which is indirectly linked to biofilm regulation in *B. cereus* since its action in exerted by binding the RNA polymerase (RNAP) core enzyme and directing it to specific promoter sequences to initiate transcription. Sigma 54 (RpoN) is involved in the transcription of virulence factors, *eps* genes and flagellins (Hayrapetyan et al., 2015).

Next to the differences mentioned above, *B. subtilis* and *B. cereus* also differ on the role of AI-2 autoinducer for the regulation of biofilm formation, such that, AI-2 autoinducer represses biofilm formation in *B. cereus* (Auger et al., 2006), while it induces biofilm formation in *B. subtilis* (Duanis-Assaf et al., 2015; Cairns et al., 2014). Furthermore, exogenous AI-2 also promotes the disintegration of cells from already formed mature biofilms (Auger et al., 2006). A variation is also about the extracellular DNA (eDNA) which is a part of extracellular polymeric matrix. eDNA in the *B. cereus* biofilm matrix is produced specifically in biofilms and it is required for adhesion on different surfaces (Vilain et al., 2009). The origin is unknown, but it is suggested to be associated with programmed cell death (Abee et al., 2011). On the other hand, in *B. subtilis*, production of eDNA requires the expression of competent genes and Opp permease and plays a role in horizontal gene transfer (Zafra et al., 2012). The pellicle morphology of *B. cereus* also differs from that of *B. subtilis*. While *B. subtilis* forms pellets with a wrinkled structure, *B. cereus* forms flat pellets (Fagerlund et al., 2014). Furthermore, when cultured, *B. subtilis* pellicles do not attach to the edges of culture tools, *B. cereus* pellicles adhere strongly to the sides of culture tools (Kobayashi, 2007a; Fagerlund et al., 2014).

GTP homeostasis and nucleotide signaling are also proposed to be important for biofilm formation in *B. cereus*. Genome-wide transcriptome analyses suggest that during biofilm formation two distinct metabolic shifts might occur, one leading to fermentation and, and the other possibly leading to production of ppGpp and c-di-GMP (Yan et al., 2017). The role of c-di-GMP on biofilm formation in *Bacillus subtilis* can be summarized such that it is reported to regulate swarming motility while having possibly very limited effect on biofilm formation (Chen et al., 2012; Gao et al., 2013). On the other hand, *Bacillus cereus* group c-di-GMP signaling is linked to biofilm formation. In contrast to *B. subtilis,* the majority of *B. cereus sensu lato* bacteria bear a set of ten genes (*cdgA-J*) encoding for diguanylate cyclases and phosphodiesterases which are responsible for c-di-GMP metabolism. Some of these genes are crucial for effective biofilm formation in an oxygenated environment, whereas some of them are also essential for motility, enterotoxin synthesis and sporulation. It should be noted that only two *cdg* genes (*cdgA*, *cdgJ*) had orthologs in *B. subtilis*, hence c-di-GMP signaling is by essence different between *B. subtilis* and *B. cereus* sensu lato (Fagerlund et al., 2016).

The other metabolic shift that should be mentioned is fermentation whose products such as acetoin and ethanol are also suggested to stimulate biofilm formation in *B. cereus*. These small volatile chemicals may directly regulate biofilm by acting as ligands or indirectly by altering expression of genes that are related to biofilm formation (Yan et al., 2017). Transcriptomic analysis reveals that *alsS* and *aldC* genes which are parts of acetoin production pathway are upregulated in biofilm culture conditions and hence, acetoin might be another biofilm enhancing signal in accordance with the studies suggesting that acetoin may enhance biofilm formation in *B. subtilis* (Yan et al., 2016; Yan et al., 2017).

Unfortunately, there is still a small amount of data describing the importance of the biofilm lifestyle in *B. anthracis*, even though biofilms are well characterized in other members of *Bacillus* species, including *Bacillus cereus* and *Bacillus subtilis*. *B. anthracis* does not need to produce biofilms for their infective cycle. Spore of *B. anthracis* is its infection agent and its toxins are highly pathogenic. Moreover, a capsule surrounds the bacteria that protects it against host immunosuppression. After the eradication of the infected host, *B. anthracis* multiplies and sporulates at the eradication site. *B. anthracis* may also preferentially produce biofilms under still and flow conditions (Lee et al., 2007; Schuch and Fischetti, 2009) and these biofilms are naturally resistant to widely used antibiotics (Lee et al., 2007). In *B. anthracis*, sporulation is also regulated during biofilm maturation while the cells encounter nutrient limitation and other stresses (Shi et al., 2004). Furthermore, although lysogeny can block or enhance sporulation, it can also induce EPS expression and biofilm formation in *B. anthracis* (Schuch and Fischetti, 2009). Despite the fact that *B. anthracis* expresses the regulators which are commonly required for biofilm formation and some of the proteinaceous components of the biofilm matrix (Pflughoeft et al., 2011), still, the contribution of biofilms to the infective life cycle of this specific bacterium is more obscure.

SinIR function and sin regulon is already discussed above for *B. subtilis* and *B. cereus.* Albeit, this regulon is well clarified in *B. subtilis* and partly in *B. cereus*, there are few reports concerning the SinIR regulatory system in *B. anthracis*, correspondingly. The *sinIR* locus in *B. anthracis* does not carry all the properties associated with its ortholog in *B. subtilis* (Mironczuk et al., 2008; Schuch and Fischetti, 2009). In *B. subtilis*, the Sin regulon is composed of many genes and approximately half of the genes controlled by SinR in each species do not have homologs in *B. anthracis*. Strikingly, it is reported that there are four genes which are commonly regulated by SinR in both species (*tasA, sipW, spoIIE*, and *spoIIG*). The *tasA* gene is located downstream of *sipW* in *B. anthracis* and unlike other members off *Bacillus spp.*, the *sinR* regulatory pathway is more appropriately adapted to the secretion of metalloproteases and this demonstrates that TasA may have a different function in *B. anthracis* (Verplaetse et al., 2015).

In *B. subtilis*, SinR represses biofilm-related *epsA-O* and *tapA–sipW–tasA* operon whilst *B. anthracis* genome does not contain similar SinR-controlled genes. Besides, *B. anthracis* is also missing the *B. subtilis* SinR-regulated *sigD* gene (encoding alternative sigma factor σ^D required for motility genes), *comS* and *srf* (competence genes), and *aprX* (encoding for a serine protease) (Pflughoeft et al., 2011). On the other hand, *B. anthracis* genes, *inhA1* and *calY* genes of the Sin regulon encode *inhA1*-secreting metalloproteases and camelysin. It should be noted that *inhA1* and *calY* genes are present in *B. cereus* too, although InhA1 was first described as a *B. thuringiensis* protease that enhances bacterial survival (Edlund et al., 1976) (Fig. 7).

The molecular signaling events start on the cell surface and result in the modulation of the specified protein cluster, leading to a variety of responses, including biofilm formation (Elsholz et al., 2014). In *B. anthracis*, PrkC is a Serine/Threonine protein kinase and the role of the sensory domain of this protein is to sense external stimulus during environmental stress conditions. Moreover, PrkC has the ability to efficiently deliver signals, although its downstream signaling cascade is not clear (Bryant-Hudson et al., 2011; Shakir et al., 2010). PrkC plays an important role in the virulence of *B. anthracis* and is crucial for survival in macrophages during infection (Arora et al., 2012; Bryant-Hudson et al., 2011; Shakir et al., 2010). Recent studies suggest that deletion of PrkC results in loss of biofilm formation and possibly cell-cell adhesion skills. PrkC is further proposed to regulate biofilm formation by enhancing a basic chaperone GroEL, although it is impossible to ignore the possible presence of other PrkC substrate proteins that could affect biofilm formation. In cooperation with GroEL, it is shown that PrkC and its homologs regulate a few pathways by phosphorylation of certain proteins and it is not unlikely that the defect in the biofilm formation originates from GroEL in *B. anthracis* (Arora et al., 2017).

The relatively sensitive arrangement of genes linked to density is defined for many bacterial species. The disclosed genome of *B. anthracis* contains an open reading frame designed to encode a LuxS ortholog which is required for the synthesis of the autoinducer-2. Hence, a functional LuxS ortholog in *B. anthracis* may be used to regulate density-dependent gene expression with extracellular signaling molecules. Since, AI-2 had a regulatory effect on biofilm formation in *B. subtilis*, *S. aureus* and *B. cereus*, the regulatory role of AI-2 on *B. anthracis* biofilm is expected but should be studied more extensively. Furthermore, a set of ten genes (*cdgA-J*) encoding for diguanylate cyclases and phosphodiesterases which are responsible for c-di-GMP metabolism is also conserved in *B. anthracis*. Therefore, the role c-di-GMP signaling which is linked to biofilm formation is also apparent in this particular pathogenic bacterium as well (Fagerlund et al., 2016).

As mentioned above, the transition into biofilm structure is also regulated by QS in *Staphylococcus aureus*. Alternative sigma factor B (σ^B) primarily triggers the stress responses for *S. aureus*, and it has a crucial effect on regulating biofilm formation (Mitchell et al., 2013). Along with the fact that σ^B (regulated by *rsbU*, *rsbW* and *rsbV*) modulates a general stress response, it also represses Agr system (Pane-Farre et al., 2006; Geiger and Wolz, 2014; Senn et al., 2005). Agr quorum sensing system is the central QS system of *S. aureus*. In accordance with the other QS responses in Gram-positive bacteria, Agr, is composed of a membrane-associated AgrC histidine kinase and the response regulator AgrA. Auto-inducing peptides expressed from *agrB* and *agrD* are responsible for the up-regulation of the expression of the effector molecule RNAIII (Novick and Geisinger, 2008). Two distinct promoters, promoters P2 and P3 are located on the upstream region of RNAII and RNAIII transcriptional units, respectively. From the P2 operon, the *agrB, agrD*, *agrC* and agrA genes are transcribed. On activation, AgrC-dependent phosphorylation results in the phosphorylation of AgrA which in turn activates P2 and P3 operons (Yarwood et al., 2004). The P2 operon induces the initiation of QS signaling cascade and P3 operon activates RNAIII expression which leads to biofilm maturation and dispersal (Karatan and Watnick, 2009; Yarwood et al., 2004). Furthermore, a *sigB*-deficient *S. aureus* cannot readily form biofilm and this deficiency contributes to the upregulation of RNAIII, which is the crucial effector of the biofilm limiting *agr* system (Rachid et al., 2000). Moreover, the deletion of *agr* or use of protease inhibitors can revert *S. aureus* from a non-biofilm forming bacterium to a biofilm-positive phenotype (Cassat et al., 2006).

Additionally, AgrA induces transcription of the alpha- and beta- phenol-soluble modulins (PSM) operon (*psmα* and *psmβ*) (Queck et al., 2008) Fig. 4). It is important to emphasize the bimodal effect of PSMs on biofilm formation. If found in a monomeric state, PSMs display surfactant properties that result in biofilm dispersal. On the other hand, when they polymerize into amyloid-like fibers then they enhance biofilm development (Dastgheyb et al., 2015; Solano et al., 2014). Recent evidences suggest that even small amounts of AgrD signal peptide may enhance PSM to be incorporated into the growing aggregate formed by self-aggregating AgrD. Moreover, the N-terminal domain of AgrD (N-AgrD) is also proposed to be a part of amyloid fibers in *S. aureus* and next to its contribution to the biofilm composition, following its cleavage with the protease SpsB, N-

AgrD also directs AIP to cell membrane (Schwartz et al., 2014; Kavanaugh. et al., 2007).

Despite the recognition of multiple functions of PSMs, regulatory mechanisms have been still mostly unclear. Still new studies suggest that MgrA can be specifically bound to the promoter regions of the *psm* operons and biofilm formation is proposed to increase significantly in the MgrA mutant strain. Hence, it is claimed that MgrA is a negative regulator of *psm* operon (Jiang et al., 2018).

Although Agr is the central QS system of *S. aureus*, recent studies also suggest the involvement of other two-component regulatory systems in biofilm formation. Programmed cell lysis in *S. aureus* is shown to happen as a response to disturbed cellular respiration. This phenomenon leads to the release of cellular polymers from which biofilms are structured. Programmed cell lysis is known to be related with murein hydrolase encoded from the *atlA* gene which is regulated by the SrrAB two-component *staphylococcal* respiratory response regulatory system (Mashruwala et al., 2017b). Along with SrrAB, biofilm formation is under the control of another two-component regulatory system SaeRS which is responsible for fermentative biofilm formation by inducing AtlA activity, and by regulating the contribution of fibronectin-binding protein A (FnBPA) into fermentative biofilm formation (Mashruwala et al., 2017a; Olson et al., 2013). Moreover, when exposed to anaerobic conditions SrrAB causes an increase in the expression of the *icaADBC* gene cluster, PIA production and consequently, biofilm formation (Mashruwala et al., 2017b) (Fig. 4).

It is important to note that *agr* activation downregulates the *icaADBC* operon (indirectly) which is required to express the extracellular polysaccharide matrix that protects the cells in the biofilm structure. Polysaccharide intercellular adhesin (PIA), also known as PNAG-poly-N-acetyl-β-(1–6)-glucosamine, is the main exopolysaccharide component of Staphylococcal biofilm matrix, thus, an important component of *S. aureus* biofilms. PIA/PNAG also renders the organisms resistant to phagocytosis. The proteins IcaA, IcaD, IcaB, and IcaC are encoded from the *ica* operon and synthesize PIA/PNAG (Cramton et al., 2001).

Regulatory effect of IcaR (intercellular adhesin locus regulator) should be considered as well since this protein can bind to a 42-bp region immediately upstream of *icaA*, by sterically hindering the binding of σ^B, thereby preventing the activation of the *ica* operon (Jefferson et al., 2003). Furthermore, AI-2 synthase gene *luxS* in *S. aureus* is also important for biofilm formation, since AI-2 can decrease biofilm formation in *S. aureus* via *icaR,* which is the major inhibitor of PIA/PNAG exopolysaccharide synthesis (Yu et al., 2012). Likewise, TcaR (teicoplanin-associated locus regulator) also negatively regulates *icaADBC, albeit weakly* (Jefferson et al., 2004) (Fig. 4).

Another positive regulator of *icaADBC* and PIA/PNAG production in *S. aureus* is Rbf, which is among the members of AraC/XylS family of transcriptional regulators (Schleif, 2010). Mode of action of Rbf is proposed to be partly through inhibiting expression of *icaR*. Next to *icaR*, Rbf also regulates a member of the Sar family of transcriptional regulatory proteins, SarX. Interestingly, *sarX* is located downstream of *rbf* in the *S. aureus* genome (Cheung et al., 2008) (Fig. 4). While Rbf can bind to *sarX* promoter, SarX can as well regulate *rbf* expression by binding to *icaA* promoter (Cue et al., 2013). Therefore, it is safe to claim that Rbf and SarX form a regulatory loop that promotes PIA-dependent biofilm formation in *S. aureus*.

It is possible to demonstrate Staphylococcal biofilm formation in two different ways: in an *ica*-dependent manner, relying upon PIA or in an *ica*-independent manner. In *ica*-independent biofilms, auxiliary proteins are responsible for the biofilm development, and this independent manner appears to be more common in methicillin resistant lineages (O'Neill et al., 2007). Cell-wall-anchored (CWA) proteins and secreted proteins together, they enhance biofilm formation. Among the 24 different CWAs which are crucial for *S. aureus* to attach to a surface and consequently fire up biofilm formation, the most extensively studied proteins are the biofilm-associated protein (Bap) (Cucarella et al., 2001), the fibronectin-binding proteins A and B (FnBPA/B) (O'Neill et al., 2008), the clumping factor B (ClfB) (Abraham and Jefferson 2012), the serine-aspartate repeat protein C

(SdrC) (Barbu et al., 2014), protein A – SpA (Merino et al., 2009) and the *S. aureus* surface protein C and G (SasC, SasG) (Schroeder et al., 2009; Geoghegan et al. ,2010). The eDNA component of biofilm matrix is also very strikingly controlled in an *ica*-independent matter, such that another auxiliary protein, Esp protease can halt biofilm formation by cleaving murein hydrolase autolysin (Atl) and this cleaving effect can stop the release of eDNA into the extracellular matrix (Chen et al., 2013; Kiedrowski et al., 2011; Okshevsky and Meyer, 2015).

ATP-dependent protease ClpP is also important for the biofilm regulatory system in *S. aureus* (Dalbey et al., 2012). While the ATPase function of Clp proteases are determined by ClpB, ClpC, ClpL and ClpX, proteolysis is provided by two hexameric rings of ClpP subunits whose activity is very important in stress response and virulence. ClpX and ClpC are known to enhance biofilm formation, whereas ClpP suppresses it (Frees et al., 2004). It is now known that the biofilm inhibitory effect of ClpP is maintained among different *S. aureus* strains. Moreover, its inhibitory effect is proposed to be at least partly because of the effect of ClpP on the Agr, as well as on the cell wall hydrolyzing enzyme Sle1 (Liu et al., 2017). Moreover, global regulator of stress response genes, SpX acts as an inhibitor for *clpP* and *clpX* (Zuber, 2004). Interestingly, Spx also inhibits biofilm formation in *S. aureus*, which is probably because of its enhancing effect on the transcription of *icaR* (Pamp et al., 2006) (Fig. 4).

It should be noted that the core genome of *S. aureus* encodes at least 16 two-component regulatory systems (TCRSs). Along with the above mentioned TCRSs, there are also evidences that show the involvement of WalKR and GraSR (Glycopeptide Resistance Associated) TCRSs in biofilm formation and regulation in *S. aureus*. Notably, the WalKR TCRS system is crucial for the maintenance of cell viability in *S. aureus*, suggesting that the organism can respond to environmental cues as well as to native bacterial factors via WALKR system (Dubrac et al., 2008). Furthermore, additional proteins like SdpC is involved in the regulation of WalKR system in *S. aureus* by exerting an inhibitory effect not only on WalK histidine kinase activity, but also on the expression of the genes that belong to WalKR regulon (Poupel et al., 2018). Next to its role in the localization of surface proteins, SpdC also has a part as an enhancer on biofilm formation and virulence. GraSR TCRS system, which is also known as ApsSR (Antimicrobial Peptide Sensor), has an important role on controlling host innate resistance against cationic antimicrobial peptides (CAMPs) (Li et al., 2007). When its expression is elevated, *S. aureus* is shown to gain increased vancomycin resistant phenotype. Moreover, controlled mutagenesis on the *graRS* operon results in CAMP sensitivity of certain *S. aureus* strains (Sass and Bierbaum, 2009) , and these studies also suggest that GraSR TCRS system also takes a part in biofilm formation (Shanks et al., 2008; Boles et al., 2010).

Biofilm Dispersal

Natural cycle of biofilm formation involves the attachment of planktonic cells to a surface, either a solid one or air-liquid interface, followed by the multiplication of the cells. Final cue comes from the environmental signals which trigger the dispersal of the biofilm structure. Throughout the dispersal process, individual cells start to regain their planktonic and motile statue and they move to new sites where another cycle of biofilm formation can be initiated. The bacterial biofilm dispersal can be categorized into three steps, where cells get separated from the biofilm mass, then transfer themselves to a new location and they finally attach to a new surface (Verstraeten et al., 2008).

Although biofilms offer many advantages to the residing cells, after the threshold biomass is reached, dispersal becomes mandatory since the matrix production brings a synthetic load on bacteria and nutrient depletion within biofilms can limit further bacterial division. Moreover, when environmental conditions become more and more unfavorable, being physically limited to the matrix turns into a burden which should be avoided, and dispersal gets initiated. It should be noted that, dispersal is clinically quite important, especially for the biofilm forming pathogens like *S. aureus*, since planktonic cells lose their resistance against antibiotic treatment when they are no longer part of the biofilm structure (Kaplan, 2010) (Fig. 8). Thus, a clear understanding of the molecular biofilm dispersal mechanism of *S. aureus* and the environmental conditions contributing to the return of these organisms into planktonic state can lead to useful therapeutic options.

Although quorum sensing mechanisms are initially effective during biofilm development, they are also efficient during dispersal. In *S. aureus*, Agr QS system has an impressive dual role, such that it is both a biofilm limiting and dispersal enhancing system which upregulates the expression of peptidases and PSMs, promoting biofilm dispersal from mature biofilms (Boles and Horswill, 2008; Lauderdale et al., 2010) (Fig. 8).

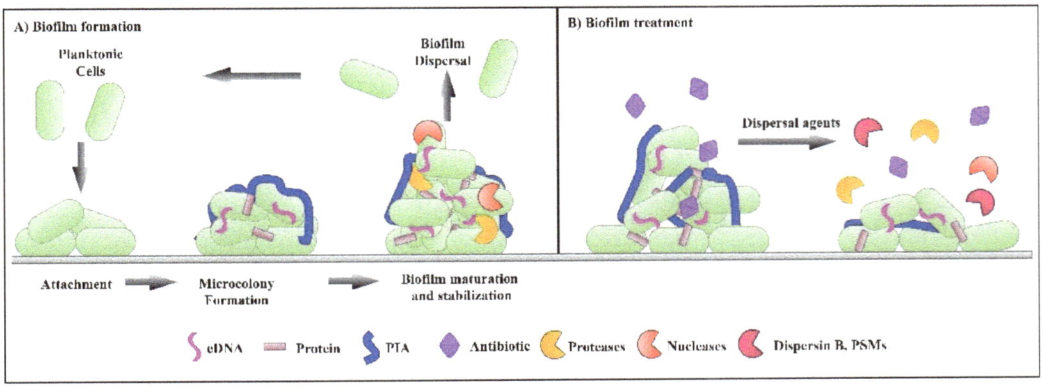

Figure 8. A) Schematic representation of the biofilm formation in *Staphylococcus aureus*. Planktonic cells attach to the surface with the help of cell wall associated proteins, which is followed by cell division and production of extracellular matrix, resulting in microcolony formation. Ongoing cell division ends up with biomass formation and biofilm maturation. Dispersal mechanisms stimulated by environmental signals result in detachment of individual cells, which remain as planktonic cells until they attach to another surface. **B)** Treatment strategies of *S. aureus* biofilms. Planktonic cells which are susceptible to antibiotics and cells close to biofilm surface that are metabolically active can be killed via antibiotic exposure. On the other hand, metabolically inactive cells located at subsurface of biofilm and resistant cells are not affected from antibiotic exposure and remain protected with the help of biofilm matrix. Exposure of biofilms with dispersal agents result in enhancement of antibiotic penetration efficiency following matrix degradation; as a result, surviving antibiotic resistant cells become susceptible to the immune system.

When the cyclic autoinducing peptide (AIP) concentration reaches a threshold level, Agr system becomes activated through a signal relay system as reviewed before and further evidences demonstrate that the *S. aureus* Agr system is enhanced in cells that have dispersed from a biofilm. Response regulator AgrA controls the expression of the alpha- and beta- phenol-soluble modulins (PSM) operon encoding PSM surfactants responsible for biofilm dispersal (Periasamy et al., 2012). PSMs take part not only in biofilm dispersal but also, they affect the volume, bulkiness and channel formation of biofilm (Dastgheyb et al., 2015; Solano et al., 2014). PSMs show surfactant-like properties with their amphiphilic alpha-helical structure and *psm* operon (including *psmα* and *psmβ*) encodes a smaller (20-25aa) PSMα, including the similar δ-toxin, and a larger PSMβ (about 45aa) (Chatterjee et al., 2013; Joo et al., 2016) Along with PSMs, extracellular proteases take a significant part in the biofilm dispersal mechanism (Fig. 8).

In *S. aureus*, both deletion of genes encoding proteases and protease inhibitor additions are shown to increase biofilm formation and create a simultaneous deficiency in dispersal following *agr* activation (Boles et al., 2010). Furthermore, expectedly, mutations on genes like *sarA* and *sigB* result in the enhancement of extracellular protease production and these mutations keep *S. aureus* from forming biofilm (Tsang et al., 2008). Thus, this finding supports the inverse correlation between protease expression and biofilm induction. Expression of the *sarA* gene results in the development of an immature biofilm, leading to the downregulation of proteases and thermostable nucleases while the expression of *sigB* reduces protease production in a similar way but additionally promotes the expression of adhesion factors which assist in the formation of the initial biofilm. Another TCRS member *arlRS*, which is a member of the OmpR-PhoB family of response modifiers, is regulated by *agr* and *sarA* locus in a way that when upregulated, the *arlS* product inhibits biofilm formation and mediates adhesion to polymer surfaces (Fournier et al., 2001). Strikingly, the *S. aureus* global regulators *sarA* and *agr* take role in a complementary and antagonistic manner. An activation cascade gets initiated first when *sarA* is induced and in turn enhances early biofilm formation by inhibiting proteases and nucleases until AIP concentration reaches a critical level (Valle et al., 2003). Once the threshold level is reached, *agr* expression antagonize *sarA* expression, leading to an increase in a number of virulence factors to inhibit the host immune response and to support the seeding dispersion. Moreover, hemolysin, enterotoxin B, serine proteases (SspA and SplA-F), cysteine proteases (SspB and ScpA), metalloprotease aureolysine (Aur), staphopains and leukotoxin D are negatively regulated, while clumping factor, fibronectin binding protein A (FnBPA) and coagulase are positively regulated by σ^B, exerting an additional control level on biofilm formation (Archer et al., 2011). The serine protease, SspA is responsible for degrading FnBPs and Bap (O'Neill et al., 2008), and Aur can degrade ClfB and Bap to modulate biofilm disruption (Abraham and Jefferson, 2012) (Fig. 4). While the staphopains can disrupt the biofilm matrix, no target proteins have been characterized yet (Mootz et al., 2013). Furthermore, AIP can also mediate the spread of the cells from the biofilm by upregulating the expression of matrix-modifying proteases, Aur and Spl (Boles and Horswill, 2008). Another regulatory protein, Rot, serves as an inhibitor for all the secreted protease genes, particularly the cysteine protease encoding genes. Therefore, *S. aureus* benefits from the transcription factor Rot to suppress secreted protease concentration in order to boost biofilm formation.

In addition to self-produced proteases, it is also revealed that the secreted proteases from different species may affect the dispersal of *S. aureus* biofilm. Interspecific matrix-degrading enzymes is discussed to possibly evolve as a bacterial nutrient source or as a defense mechanism that pushes and displaces competing species. For instance, the serine protease ESP, produced by *Staphylococcus epidermidis* is responsible for the dispersal from *S. aureus* biofilms (Sugimoto et al., 2013). The role attributed to ESP is to digest several *S. aureus* proteins including Eap, FnBPA and Atl (Chen et al., 2013; Sugimoto et al., 2013). The mechanism of Esp-mediated dispersal is accomplished in two ways: First, Esp targets matrix proteins that are important for intercellular adhesion and secondly it targets autolysin Atl, a murein hydrolase, which results in the release of eDNA to the extracellular environment (Iwase et al., 2010; Sugimoto et al., 2013, Chen et al., 2013). An orthologous Cid/Lrg system was shown to be responsible for cell death and DNA release in *S. aureus* biofilms, although the role of this system in biofilm dispersal should be further investigated (Rice et al., 2007; Mann et al., 2009). The *cidA* gene is involved in cell lysis, which is essential for the release of genomic DNA and biofilm development (Rice et al., 2007). The *lrg* gene, on the other hand, acts antagonistically to *cidA* by canceling out the effects of DNA discharge into extracellular space (Mann et al., 2009). The release of dead cells and genomic DNA from a *S. aureus* biofilm is functionally similar to the role of apoptosis in eukaryotes. Some studies suggest that the Cid/Lrg system is involved in the activation of *S. aureus* programmed cell death (PCD), and it is not known how the PCD is induced only in a subpopulation of cells, although it is an interesting question that should be further addressed to be able to understand differentiation in bacterial biofilms as well. It should be emphasized that exploitation of these systems may help to identify matrix degrading enzymes that regulate dispersal of biofilms (Karatan and Watnick, 2009; Landini et al., 2010).

As the biofilm matrix covers the bacterial cells in the biofilm, disruption of the matrix results in separation of the cells and their release into the environment. The matrix degrading enzymes involved in the active biofilm spread include glycosidases, proteases and deoxyribonucleases. Expectedly, Bovine DNAse I supplementation into the medium is also shown to be successful in the disintegration of *S. aureus* biofilms (Mann et al., 2009; Izano et al., 2008a). Such experiments have contributed to the evidence that not only proteins or carbohydrates but also eDNA is an important structural component of the *staphylococcal* biofilm matrix. Clinically, recombinant form of DNase I (pulmozyme) has the ability to degrade eDNA and consequently, it can be used against *staphylococcal* infections (Shak, et al., 1990; Fey, 2010). Although there are many reports claiming that DNase I or restriction enzymes can easily disrupt bacterial biofilms (Webb et al., 2003; Izano et al., 2008a; Mann et al., 2009; Kaplan et al., 2003), still there are a few studies reporting the production of nucleases in *S. aureus* as a means of biofilm disruption. Deoxyribonuclease (thermonuclease and micrococcal nuclease) is expectantly involved in cell detachment in *S. aureus* biofilms (Mann et al., 2009). In fact, in *S. aureus* these extracellular nucleases are encoded from *nuc* and surface attached nuclease Nuc2 is encoded from *nuc2* (Tang et al., 2008; Kiedrowski et al., 2014; Lister and Horswill, 2014). The nuclease production is regulated by *sigB* and SaeRS two-component system (Kiedrowski et al., 2011; Olson et al., 2013) and *nuc* expression is reported to be very low during biofilm growth conditions (Olson et al., 2013). Moreover, *nuc* is reported to be transcriptionally upregulated in *sarA* mutants along with extracellular proteases. Next to its role on modulating biofilm, Nuc is also responsible for the disruption of neutrophil extracellular traps (NETs) during infection. The endogenous role of Nuc2 is not as clear as the role of Nuc, whereas Nuc2 supplementation partially contributes to the dispersal of existing biofilms, indicating that Nuc2 might be involved in localized dispersal via creating hollow cavities that reach internal parts of biofilm colonies or via supporting Nuc activity in high flow environments where Nuc cannot resist against flow (Kiedrowski et al., 2014).

Furthermore, biofilm formation and dispersal of *S. aureus* can also be controlled by oxacillin which is used in resistant *staphylococci* infections. It is proposed that oxacillin has a regulatory effect on *icaA* expression, inducing *icaA*-dependent PIA production and biofilm formation. It is also proposed that once oxacillin starts to lose its toxic effect, AgrA becomes activated leading to the release of the planktonic cells from biofilm colonies (Mirani et al., 2013). Moreover, when supplemented at low concentration, ascorbic acid also inhibits EPS production and biofilm formation in *S. aureus*. The inhibition of biofilm formation is enhanced when the cells are treated with only ascorbic acid or together with oxacillin. Furthermore, transcriptomic studies further prove that ascorbic acid upregulates *agr* expression and downregulates *icaA* gene expression as mentioned in the case of oxacillin (Mirani et al., 2018).

Biofilm matrix components are main targets for biofilm degradation. As it mentioned above, various enzymes can be used in the process of biofilm dispersal. Proteinase K and trypsin are among those enzymes that are shown to disrupt biofilms produced by *S. aureus* (Mootz et al., 2013). Dispersin B (DspB), which is a β-hexosaminidase produced by *Actinobacillus actinomycetem-comitans*, hydrolyzes glyosidic bonds, and therefore, it can disperse biofilms which contain PIA. This property of DspB makes it a strong candidate as an anti-biofilm molecule, however since it is possible to encounter *S. aureus* strains which are not dependent on polysaccharides, usage of DspB against infections would have definite limitations. There are no genes homologous to *dspB* in *S. aureus* genome, but external supplementation of DspB decreases the resistance of *S. aureus* against antimicrobial agents which signals the effect of DspB on *S. aureus* biofilm dispersal (Donelli et al., 2007). Furthermore, iron restriction has an enhancing effect on DspB levels, promoting bacteria to avoid iron-deficient environment (Stacy et al., 2016). Genes homologous to *dspB* have been discovered in other bacterial genomes, but DspB-like activity of the proteins are not known at the moment (Kaplan et al., 2004; Chaignon et al., 2007; Izano et al., 2008b). Moreover, complementary usage of antiseptic agents together with DspB shows promising results as anti-biofilm/antimicrobial treatment against *S. aureus, S. epidermidis* and *E. coli* (Darouiche et al., 2009). Therefore, DspB may also disrupt biofilms that are produced by a variety of different species such as those encountered in oral cavities. Finally, lysostaphin which is a glycine endopeptidase produced by *Staphylococcus simulans* disrupts the maturated biofilms of *S. aureus* on abiotic surfaces by cleaving the pentaglycan bridge in the cell wall. Under the light of this information, lysostaphin is proposed to be useful as a collective treatment option for *staphylococcal* biofilm infections (Kokai-Kun et al., 2009).

Another phenomenon also called as seeding dispersal involves mature biofilms which contain hollow cavities that reach to the central zone inside the biofilm. These cavities have been observed in biofilms produced by *S. aureus* (Kaplan et al., 2003; Yarwood et al., 2004; Koh et al., 2007; Ma et al., 2009) and they play a role in biofilm cell detachment through seeding dispersal which happens when a gap in the extracellular wall lets the passage of the planktonic cells from the cavities. Although these cavities are not extensively studies, they seem to be involved with the cells that are perished inside the colony.

Along with the foreign factors secreted by other organisms, changes in the environmental conditions also result in *S. aureus* biofilm dispersal. These environmental cues consist of availability of nutrients, oxygen saturation, and variation in cell biomass. For instance, under normal biofilm growth conditions, glucose suppresses the *agr* system by creating a low pH level with the help of excreted acidic metabolites and thus, glucose is generally used to boost *staphylococcal* biofilm formation by many laboratories. When glucose is no longer available to *S. aureus*, *agr* system becomes reactivated and the production of matrix-degrading enzymes and surfactants results in dispersal of the biofilm (McDougald et al., 2011; Solano et al., 2014; Kim and Lee, 2016).

Other communication molecules that have a role in the dispersal of *B. subtilis* and *S. aureus* biofilm cells also include the fatty acid cis-2-decenoic acid (CDA), produced by *Pseudomonas aeruginosa*. When found at low concentrations, CDA is proposed to function as a cell-cell communication molecule in bacteria and fungi, considering the observations that CDA resembles to the class of short chain fatty acid signaling molecules. Combined treatments of antibiotics or disinfectants with CDA has a great promise for the biofilm control of these organisms (Sepehr et al., 2014). Moreover, autoinducers 2 (AI-2), which is used for internal and interspecific communication, is also thought to be involved in biofilm dispersal in a variety of bacteria including *S. aureus* (Yu et al., 2012). Nevertheless, the role of AI-2-based QS regulatory system in the regulation of Staphylococci biofilm is still unclear.

D-amino acids also have an effect on biofilm dispersal in a number of bacteria including *S. aureus* (Kolodkin-Gal et al., 2010). D-amino acids are known to prevent the surface localization of proteins in this organism. The first discovery of the amino acid involvement in dispersal was in *B. subtilis*. In *B. subtilis*, the proposed mechanism of this phenomenon was based on the incorporation of D-amino acids into the peptidoglycan layer. In mature biofilms, release of D-amino acids happens naturally for many biofilm-forming bacteria. *B. subtilis* is a representative of these organisms from which a cocktail of D-amino acids that constitutes of D-tyrosine, D-leucine, D-tryptophan and D-methionine. are released and subsequently this directs bacteria into the dispersal and inhibition of further biofilm formation (Hochbaum et al., 2011). It should be noted that ECM of biofilms in *B. subtilis* partly consists of amyloid fibers that are crucial for the integrity of the biofilm and this mixture of D-amino acids are responsible for releasing these amyloid fibers (Romero et al., 2011). The D-amino acids found in the mixture replace D-alanine which is normally located on the peptidoglycan peptide side chain and hence this replacement causes the release of TapA which eventually leads to the release of amyloid fibers from the cell (Romero et al., 2011; Driks, 2011). Additionally, racemase enzymes that convert L-form amino acids into D-form amino acids are shown to regulate biofilm dispersal in a possibly indirect way, since when racemase-encoding genes are mutated, significantly late-appearing dispersal can be observed. D-amino acids also prevent biofilm formation in *S. aureus* although this bacterium has no TapA homologue (Hochbaum et al., 2011). Hence, it is suggested that D-amino acids might be a common component responsible for biofilm dispersal among species (Kolodkin-Gal et al., 2010). This mechanism is debatable though, since another study claims that, when mutated, *dtd* gene (encodes D-tyrosyl- tRNA deacylase) becomes responsible for the misincorporation of D-amino acids into proteins. This data signals possibly indirect consequence of their toxic effect on bacterial dispersal through protein synthesis (Leiman et al., 2013). Moreover, showing amyloid-degrading capabilities, protease nattokinase of *B. subtilis Natto* has a role in the release of TasA amyloid-like fibres (Hsu et al., 2009).

The norspermidine is postulated to be a naturally produced polyamine and also a strong inhibitor of biofilm formation in *B. subtilis* and *S. aureus*. On the other hand, biofilm-inhibiting properties of norspermidine in *S. aureus* need more clarification (Kolodkin-Gal et al., 2010). In contrast to the one of the possible activities of D-amino acids, norspermidine is proposed not to exert its inhibitory effect on TasA and amyloid fibers, instead, it directly deteriorates the EPS component of the matrix. Not only norspermidine, but also norspermine is suggested to efficiently disperse biofilms in *B. subtilis*. However, more recent studies have shown that *B. subtilis* does not harbor norspermidine biosynthetic pathway (Hobley et al., 2014). Hence, the norspermidine-dependent dispersal mechanism of *B. subtilis* biofilms demands more elaborated studies. Along with these agents, surfactin in *B. subtilis* is also the potent surfactant molecule that promote biofilm dispersal (Kinsinger et al., 2003; Kinsinger et al., 2005; Angelini et al., 2009). Additionally, rhamnolipids are the interspecies-regulatory agents that act on *B. subtilis* dispersal, just like fatty acid cis-2-decenoic acid (CDA), produced by *Pseudomonas aeruginosa*. When synthesized by *Pseudomonas aeruginosa*, rhamnolipids can in turn act directly on the biofilm matrix of other species like *Bordetella bronchiseptica*, *B. subtilis* and *S. epidermidis* (Irie et al., 2005; Pihl et al., 2013; De Rienzo and Martin, 2016; Pamp and Tolker-Nielsen, 2007). Moreover, CDA also activates biofilm dispersal, formed by *Bacillus cereus*, although the mechanism is not enlightened yet (Sepehr et al., 2014). *B. cereus* and *B. anthracis* biofilm dispersal mechanisms are scarcely investigated, and further data should accumulate for a healthier review.

Concluding remarks

Biofilm formation and dispersal is under the control of complex regulatory mechanisms that involve the coordination of various signaling messengers and molecular effectors. Like biofilm formation, biofilm dispersal is also a critical step in bacterial lifestyle allowing dissemination of bacteria from the original site of infection in the host or the environment and it could increase virulence of pathogenic bacteria. Advances in our understanding in the regulation of biofilm formation and dispersal of various bacteria could lead to the development of clinically useful agents that inhibit biofilm formation and/or promote biofilm detachment within the industrial and healthcare settings, even though there is no single mechanism of biofilm formation and dispersal used by all bacteria. In this regard, finding a treatment to interrupt the complex regulatory systems involved in biofilm formation is a promising strategy. Additionally, combinatorial therapy including the combined treatment of antibiotics with biofilm-dispersal agents have been shown to provide a very promising strategy for the treatment of biofilm-related infections. To date, an exciting amount of potential dispersal-inducing agents have been identified. On the other hand, most of these dispersal agents were tested *in-vitro* with monospecies biofilms under the controlled-conditions of a laboratory. It is extremely difficult to extrapolate these results to any complex multispecies environmental biofilms. Therefore, we need more *in-vivo* experiments and we need to further assess whether any of these agents will have clinical significance or not.

References

Abee, T., Kovacs, A.T., Kuipers, O.P., and van der Veen, S. (2011). Biofilm formation and dispersal in Gram-positive bacteria. Curr Opin Biotech 22, 172-179. http://doi:110.1016/j.copbio. 2010.1010.1016

Abraham, N.M., and Jefferson, K.K. (2012). *Staphylococcus aureus* clumping factor B mediates biofilm formation in the absence of calcium. Microbiol-Sgm 158, 1504-1512. https://doi:10.1099/mic.0.057018-0

Agaisse, H., Gominet, M., Okstad, O.A., Kolsto, A.B., and Lereclus, D. (1999). PlcR is a pleiotropic regulator of extracellular virulence factor gene expression in *Bacillus thuringiensis*. Mol Microbiol 32, 1043-1053. http://dx.doi.org/1010.1046/j.1365-2958.1999.01419.x

Aguilar, C., Vlamakis, H., Guzman, A., Losick, R., and Kolter, R. (2010). KinD is a checkpoint protein linking spore formation to extracellular-matrix production in *Bacillus subtilis* biofilms. MBio 1, e00035-10. http://doi:10.1128/mBio.00035-10

Angelini, T.E., Roper, M., Kolter, R., Weitz, D.A., and Brenner, M.P. (2009). *Bacillus subtilis* spreads by surfing on waves of surfactant. Proc Natl Acad Sci USA 106, 18109-18113. http://dx.doi.org/18110.11073/pnas.0905890106

Ansaldi, M., Marolt, D., Stebe, T., Mandic-Mulec, I., and Dubnau, D. (2002). Specific activation of the *Bacillus* quorum-sensing systems by isoprenylated pheromone variants. Mol Microbiol 44, 1561-1573. http://dx.doi.org/1510.1046/j.1365-2958.2002.02977.x

Archer, N.K., Mazaitis, M.J., Costerton, J.W., Leid, J.G., Powers, M.E., and Shirtliff, M.E. (2011). *Staphylococcus aureus* biofilms Properties, regulation and roles in human disease. Virulence 2, 445-459. https://doi:10.4161/viru.2.5.17724

Arora, G., Sajid, A., Arulanandh, M.D., Singhal, A., Mattoo, A.R., Pomerantsev, A.P., Leppla, S.H., Maiti, S., and Singh, Y. (2012). Unveiling the Novel Dual Specificity Protein Kinases in *Bacillus anthracis*. J Biol Chem 287, 26749-26763. http://doi:10.1074/jbc.M112.351304

Arora, G., Sajid, A., Virmani, R., Singhal, A., Kumar, C.M.S., Dhasmana, N., Khanna, T., Maji, A., Misra, R., Molle, V., et al. (2017). Ser/Thr protein kinase PrkC-mediated regulation of GroEL is critical for biofilm formation in *Bacillus anthracis*. Npj Biofilms Microbi 3, 7. http://doi:10.1038/s41522-017-0015-4

Asally, M., Kittisopikul, M., Rue, P., Du, Y., Hu, Z., Cagatay, T., Robinson, A.B., Lu, H., Garcia-Ojalvo, J., and Suel, G.M. (2012). Localized cell death focuses mechanical forces during 3D patterning in a biofilm. PNAS USA 109, 18891-18896. http://dx.doi.org/18810.11073/pnas.1212429109

Auchtung, J.M., Lee, C.A., and Grossman, A.D. (2006). Modulation of the ComA-dependent quorum response in *Bacillus subtilis* by multiple Rap proteins and Phr peptides. J Bacteriol 188, 5273-5285. http://dx.doi.org/5210.1128/JB.00300-00306

Auger, S., Krin, E., Aymerich, S., and Gohar, M. (2006). Autoinducer 2 affects biofilm formation by *Bacillus cereus*. Appl Environ Microb 72, 937-941. http://doi:10.1128/AEM.72.1.937-941.2006

Bacon Schneider, K., Palmer, T.M., and Grossman, A.D. (2002). Characterization of *comQ* and *comX*, two genes required for production of ComX pheromone in *Bacillus subtilis*. J Bacteriol 184, 410-419. http://doi:10.1128/JB.184.2.410-419.2002

Bai, U., Mandic-Mulec, I., and Smith, I. (1993). SinI modulates the activity of SinR, a developmental switch protein of *Bacillus subtilis*, by protein-protein interaction. Genes Dev 7, 139-148. http://doi:10.1101/gad.7.1.139

Baker, M.D., and Neiditch, M.B. (2011). Structural basis of response regulator inhibition by a bacterial anti-activator protein. PLoS Biol 9, e1001226. http://dx.doi.org/1001210.1001371/journal.pbio.1001226

Banse, A.V., Chastanet, A., Rahn-Lee, L., Hobbs, E.C., and Losick, R. (2008). Parallel pathways of repression and antirepression governing the transition to stationary phase in *Bacillus subtilis*. Proc Natl Acad Sci USA 105, 15547-15552. http://dx.doi.org/15510.11073/pnas.0805203105

Barbu, E.M., Mackenzie, C., Foster, T.J., and HooK, M. (2014). SdrC induces staphylococcal biofilm formation through a homophilic interaction. Mol Microbiol 94, 172-185. https://doi:10.1111/mmi.12750

Bassler, B.L. (1999). How bacteria talk to each other: regulation of gene expression by quorum sensing. Curr Opin Microbiol 2, 582-587. http://dx.doi.org/510.1016/S1369-5274(1099)00025-00029

Bazinet, A.L. (2017). Pan-genome and phylogeny of *Bacillus cereus sensu lato*. BMC Evol Biol 17, 176. http://dx.doi.org/110.1186/s12862-12017-11020-12861

Boles, B.R., and Horswill, A.R. (2008). *agr*-mediated dispersal of *Staphylococcus aureus* biofilms. Plos Pathog 4, e1000052. https://doi:10.1371/journal.ppat.1000052

Boles, B.R., Thoendel, M., Roth, A.J., and Horswill, A.R. (2010). Identification of Genes Involved in Polysaccharide-Independent *Staphylococcus aureus* Biofilm Formation. Plos One 5, e10146. https://doi:10.1371/journal.pone.0010146

Bongiorni, C., Ishikawa, S., Stephenson, S., Ogasawara, N., and Perego, M (2005). Synergistic regulation of competence development in *Bacillus subtilis* by two Rap-Phr systems. J Bacteriol 187, 4353-4361. http://dx.doi.org/4310.1128/JB.4187.4313.4353-4361.2005

Bouillaut, L., Perchat, S., Arold, S., Zorrilla, S., Slamti, L., Henry, C., Gohar, M. Declerck, N., and Lereclus, D. (2008). Molecular basis for group-specific activation of the virulence regulator PlcR by PapR heptapeptides. Nucleic Acids Res 36, 3791-3801. http://dx.doi.org/3710.1093/nar/gkn3149

Branda, S.S., Chu, F., Kearns, D.B., Losick, R., and Kolter, R. (2006). A major protein component of the *Bacillus subtilis* biofilm matrix. Mol Microbiol 59, 1229-1238. http://dx.doi.org/1210.1111/j.1365-2958.2005.05020.x

Branda, S.S., Gonzalez-Pastor, J.E., Ben-Yehuda, S., Losick, R., and Kolter, R. (2001). Fruiting body formation by *Bacillus subtilis*. Proc Natl Acad Sci USA 98, 11621-11626. http://dx.doi.org/11610.11073/pnas.191384198

Branda, S.S., Gonzalez-Pastor, J.E., Dervyn, E., Ehrlich, S.D., Losick, R., and Kolter, R. (2004). Genes involved in formation of structured multicellular communities by *Bacillus subtilis*. J Bacteriol 186, 3970-3979. http://dx.doi.org/3910.1128/JB.3186.3912.3970-3979.2004

Bryant-Hudson, K.M., Shakir, S.M., and Ballard, J.D. (2011). Autoregulatory Characteristics of a *Bacillus anthracis* Serine/Threonine Kinase. J Bacteriol 193, 1833-1842. http://doi:10.1128/JB.01401-10

Cairns, L.S., Hobley, L., and Stanley-Wall, N.R. (2014). Biofilm formation by *Bacillus subtilis*: new insights into regulatory strategies and assembly mechanisms. Mol Microbiol 93, 587-598. http://doi: 10.1111/mmi.12697

Candela, T., Maes, E., Garenaux, E., Rombouts, Y., Krzewinski, F., Gohar, M., and Guerardel, Y. (2011). Environmental and biofilm-dependent changes in a *Bacillus cereus* secondary cell wall polysaccharide. J Biol Chem 286, 31250-31262. http://doi:31210.31074/jbc.M31111.249821

Caro-Astorga, J., Perez-Garcia, A., de Vicente, A., and Romero, D. (2015). A genomic region involved in the formation of adhesin fibers in *Bacillus cereus* biofilms. Front Microbiol 5, 745. http://dx.doi.org/710.3389/fmicb.2014.00745

Cassat, J., Dunman, P.M., Murphy, E., Projan, S.J., Beenken, K.E., Palm, K.J., Yang, S.J., Rice, K.C., Bayles, K.W., and Smeltzer, M.S. (2006). Transcriptional profiling of a *Staphylococcus aureus* clinical isolate and its isogenic *agr* and *sarA* mutants reveals global differences in comparison to the laboratory strain RN6390. Microbiol-Sgm 152, 3075-3090. http://doi: 10.1099/mic.0.29033-0

Chai, Y., Beauregard, P.B., Vlamakis, H., Losick, R., and Kolter, R. (2012). Galactose metabolism plays a crucial role in biofilm formation by *Bacillus subtilis*. MBio 3, e00184-00112. http://dx.doi.org/00110.01128/mBio.00184-00112

Chai, Y., Kolter, R., and Losick, R. (2010a). Reversal of an epigenetic switch governing cell chaining in *Bacillus subtilis* by protein instability. Mol Microbiol 78, 218-229. http://dx.doi.org/210.1111/j.1365-2958.2010.07335.x

Chai, Y., Norman, T., Kolter, R., and Losick, R. (2010b). An epigenetic switch governing daughter cell separation in *Bacillus subtilis*. Genes Dev 24, 754-765. http://dx.doi.org/710.1101/gad.1915010

Chaignon, P., Sadovskaya, I., Ragunah, C., Ramasubbu, N., Kaplan, J.B., and Jabbouri, S. (2007). Susceptibility of *staphylococcal* biofilms to enzymatic treatments depends on their chemical composition. Appl Microbiol Biotechnol 75, 125-132. http://dx.doi.org/110.1007/s00253-00006-00790-y

Chan, W.C., Coyle, B.J., and Williams, P. (2004). Virulence regulation and quorum sensing in *staphylococcal* infections: competitive AgrC antagonists as quorum sensing inhibitors. J Med Chem 47, 4633-4641. http://dx.doi.org/4610.1021/jm0400754

Chatterjee, S.S., Joo, H.S., Duong, A.C., Dieringer, T.D., Tan, V.Y., Song, Y., Fischer, E.R., Cheung, G.Y.C., Li, M., and Otto, M. (2013). Essential *Staphylococcus aureus* toxin export system. Nat Med 19, 364-367. https://doi:10.1038/nm.3047

Chen, C., Krishnan, V., Macon, K., Manne, K., Narayana, S.V.L., and Schneewind, O. (2013). Secreted Proteases Control Autolysin-mediated Biofilm Growth of *Staphylococcus aureus*. J Biol Chem 288, 29440-29452. https://doi:10.1074/jbc.M113.502039

Chen, G., Kumar, A., Wyman, T.H., and Moran, C.P., Jr. (2006). Spo0A-dependent activation of an extended -10 region promoter in *Bacillus subtilis*. J Bacteriol 188, 1411-1418. http://dx.doi.org/1410.1128/JB.1188.1414.1411-1418.2006

Chen, R., Guttenplan, S.B., Blair, K.M., and Kearns, D.B. (2009). Role of the σ^D-dependent autolysins in *Bacillus subtilis* population heterogeneity. J Bacteriol 191, 5775-5784. http://dx.doi.org/5710.1128/JB.00521-00509

Chen, Y., Chai, Y., Guo, J.H., and Losick, R. (2012). Evidence for cyclic Di-GMP-mediated signaling in *Bacillus subtilis*. J Bacteriol 194, 5080-5090. http://dx.doi.org/5010.1128/JB.01092-01012

Chen, Y.P., Zhang, P., Guo, J.S., Fang, F., Gao, X., and Li, C. (2013). Functional groups characteristics of EPS in biofilm growing on different carriers. Chemosphere 92, 633-638. http://dx.doi.org/610.1016/j.chemosphere.2013.1001.1059

Cheung, A.L., Bayer, A.S., Zhang, G., Gresham, H., and Xiong, Y.Q. (2004). Regulation of virulence determinants in vitro and in vivo in *Staphylococcus aureus*. FEMS Immunol Med Microbiol 40, 1-9. http://doi:10.1016/S0928-8244(1003)00309-00302

Cheung, A.L., Nishina, K.A., Trotonda, M.P., and Tamber, S. (2008). The SarA protein family of *Staphylococcus aureus*. Int J Biochem Cell B 40, 355-361. https://doi:10.1016/j.biocel.2007.10.032

Chu, F., Kearns, D.B., Branda, S.S., Kolter, R., and Losick, R. (2006). Targets of the master regulator of biofilm formation in *Bacillus subtilis*. Mol Microbiol 59, 1216-1228. http://doi:1210.1111/j.1365-2958.2005.05019.x

Chu, F., Kearns, D.B., McLoon, A., Chai, Y., Kolter, R., and Losick, R. (2008). A novel regulatory protein governing biofilm formation in *Bacillus subtilis*. Mol Microbiol 68, 1117-1127. http://doi:1110.1111/j.1365-2958.2008.06201.x

Comella, N., and Grossman, A.D. (2005). Conservation of genes and processes controlled by the quorum response in bacteria: characterization of genes controlled by the quorum-sensing transcription factor ComA in *Bacillus subtilis*. Mol Microbiol 57, 1159-1174. http://dx.doi.org/1110.1111/j.1365-2958.2005.04749.x

Core, L., and Perego, M. (2003). TPR-mediated interaction of RapC with ComA inhibits response regulator-DNA binding for competence development in *Bacillus subtilis*. Mol Microbiol 49, 1509-1522. http://dx.doi.org/1510.1046/j.1365-2958.2003.03659.x

Costerton, J.W. (1999). Introduction to biofilm. Int J Antimicrob Agents 11, 217-221; discussion 237-219. http://dx.doi.org/210.1016/S0924-8579(1099)00018-00017

Costerton, J.W., Stewart, P.S., and Greenberg, E.P. (1999). Bacterial biofilms: a common cause of persistent infections. Science 284, 1318-1322. http://dx.doi.org/1310.1126/science. 1284.5418.1318

Cramton, S.E., Ulrich, M., Gotz, F., and Doring, G. (2001). Anaerobic conditions induce expression of polysaccharide intercellular adhesin in *Staphylococcus aureus* and *Staphylococcus epidermidis*. Infect Immun 69, 4079-4085. https://doi:10.1128/IAI.69.6.4079-4085.2001

Cucarella, C., Solano, C., Valle, J., Amorena, B., Lasa, I., and Penades, J.R. (2001). Bap, a *Staphylococcus aureus* surface protein involved in biofilm formation. J Bacteriol 183, 2888-2896. https://doi:10.1128/JB.183.9.2888-2896.2001

Cue, D., Lei, M.G., and Lee, C.Y. (2013). Activation of *sarX* by Rbf Is Required for Biofilm Formation and *icaADBC* Expression in *Staphylococcus aureus*. J Bacteriol 195, 1515-1524. https://doi:10.1128/JB.00012-13

Dalbey, R.E., Wang, P., and van Dijl, J.M. (2012). Membrane Proteases in the Bacterial Protein Secretion and Quality Control Pathway. Microbiol Mol Biol R 76, 311-330. https://doi:10.1128/MMBR.05019-11

Daniels, R., Vanderleyden, J., and Michiels, J. (2004). Quorum sensing and swarming migration in bacteria. FEMS Microbiol Rev 28, 261-289. http://dx.doi.org/210.1016/j.femsre.2003.1009.1004

Darouiche, R.O., Mansouri, M.D., Gawande, P.V., and Madhyastha, S. (2009). Antimicrobial and antibiofilm efficacy of triclosan and Dispersin B combination. J Antimicrob Chemother 64, 88-93. http://doi:10.1093/jac/dkp1158

Dastgheyb, S.S., Villaruz, A.E., Le, K.Y., Tan, V.Y., Duong, A.C., Chatterjee, S.S., Cheung, G.Y., Joo, H.S., Hickok, N.J., and Otto, M. (2015). Role of Phenol-Soluble Modulins in Formation of *Staphylococcus aureus* Biofilms in Synovial Fluid. Infect Immun 83, 2966-2975. http://doi:2910.1128/IAI.00394-00315

De Rienzo, M.A., and Martin, P.J. (2016). Effect of Mono and Di-rhamnolipids on Biofilms Pre-formed by *Bacillus subtilis* BBK006. Curr Microbiol 73, 183-189. http://doi:110.1007/s00284-00016-01046-00284

Declerck, N., Bouillaut, L., Chaix, D., Rugani, N., Slamti, L., Hoh, F., Lereclus, D., and Arold, S.T. (2007). Structure of PlcR: Insights into virulence regulation and evolution of quorum sensing in Gram-positive bacteria. Proc Natl Acad Sci USA 104, 18490-18495. http://dx.doi.org/18410.11073/pnas.0704501104

Dietrich, L.E., Teal, T.K., Price-Whelan, A., and Newman, D.K. (2008). Redox-active antibiotics control gene expression and community behavior in divergent bacteria. Science 321, 1203-1206. http://dx.doi.org/1210.1126/science.1160619

Donelli, G., Francolini, I., Romoli, D., Guaglianone, E., Piozzi, A., Ragunath, C., and Kaplan, J.B. (2007). Synergistic activity of dispersin B and cefamandole nafate in inhibition of *staphylococcal* biofilm growth on polyurethanes. Antimicrob Agents Chemother 51, 2733-2740. http://doi:2710.1128/AAC.01249-01206

Donlan, R.M., and Costerton, J.W. (2002). Biofilms: survival mechanisms of clinically relevant microorganisms. Clin Microbiol Rev 15, 167-193. http://dx.doi.org/110.1128/ CMR.1115.1122.1167-1193.2002

Driks, A. (2011). Tapping into the biofilm: insights into assembly and disassembly of a novel amyloid fibre in *Bacillus subtilis*. Mol Microbiol 80, 1133-1136. https://doi.org/1110.1111/j.1365-2958.2011.07666.x

Duanis-Assaf, D., Steinberg, D., Chai, Y.R., and Shemesth, M. (2015). The LuxS Based Quorum Sensing Governs Lactose Induced Biofilm Formation by *Bacillus subtilis*. Front Microbiol 6, 1517. http://dx.doi.org/10.3389/fmicb.2015.01517

Dubois, T., Faegri, K., Perchat, S., Lemy, C., Buisson, C., Nielsen-LeRoux, C., Gohar, M., Jacques, P., Ramarao, N., Kolsto, A.B., et al. (2012). Necrotrophism is a quorum-sensing-regulated lifestyle in *Bacillus thuringiensis*. PLoS Pathog 8, e1002629. http://doi:1002610.1001371/journal.ppat.1002629

Dubois, T., Perchat, S., Verplaetse, E., Gominet, M., Lemy, C., Aumont-Nicaise, M., Grenha, R., Nessler, S., and Lereclus, D. (2013). Activity of the *Bacillus thuringiensis* NprR-NprX cell-cell communication system is co-ordinated to the physiological stage through a complex transcriptional regulation. Mol Microbiol 88, 48-63. http://doi:10.1111/mmi.12168

Dubrac, S., Bisicchia, P., Devine, K.M., and Msadek, T. (2008). A matter of life and death: cell wall homeostasis and the WalKR (YycGF) essential signal transduction pathway. Mol Microbiol 70, 1307-1322. https://doi:10.1111/j.1365-2958.2008.06483.x

Edlund, T., Siden, I., and Boman, H.G. (1976). Evidence for two immune inhibitors from *Bacillus thuringiensis* interfering with the humoral defense system of saturniid pupae. Infection and Immunity 14, 934–941.

Elsholz, A.K.W., Wacker, S.A., and Losick, R. (2014). Self-regulation of exopolysaccharide production in *Bacillus subtilis* by a tyrosine kinase. Gene Dev 28, 1710-1720. http://doi:10.1101/gad.246397.114

Fagerlund, A., Dubois, T., Okstad, O.A., Verplaetse, E., Gilois, N., Bennaceur, I., Perchat, S., Gominet, M., Aymerich, S., Kolsto, A.B., et al. (2014). SinR controls enterotoxin expression in *Bacillus thuringiensis* biofilms. PLoS One 9, e87532. http://dx.doi.org/87510.81371/ journal.pone.0087532

Fagerlund, A., Smith, V., Rohr, A.K., Lindback, T., Parmer, M.P., Andersson, K.K., Reubsaet, L., and Okstad, O.A. (2016). Cyclic diguanylate regulation of *Bacillus cereus* group biofilm formation. Mol Microbiol 101, 471-494. http://doi:410.1111/mmi.13405

Fey, P.D. (2010). Modality of bacterial growth presents unique targets: how do we treat biofilm-mediated infections? Curr Opin Microbiol 13, 610-615. http://doi:610.1016/j.mib. 2010.1009.1007

Flemming, H.C., and Wingender, J. (2010). The biofilm matrix. Nat Rev Microbiol 8, 623-633. http://dx.doi.org/610.1038/nrmicro2415

Fournier, B., Klier, A., and Rapoport, G. (2001). The two-component system ArlS-ArlR is a regulator of virulence gene expression in *Staphylococcus aureus*. Mol Microbiol 41, 247-261. https://doi.org/10.1046/j.1365-2958.2001.02515.x

Frees, D., Chastanet, A., Qazi, S., Sorensen, K., Hill, P., Msadek, T., and Ingmer, H. (2004). Clp ATPases are required for stress tolerance, intracellular replication and biofilm formation in *Staphylococcus aureus*. Mol Microbiol 54, 1445-1462. https://doi.org/10.1111/j.1365-2958.2004.04368.x

Frenzel, E., Doll, V., Pauthner, M., Lucking, G., Scherer, S., and Ehling-Schulz, M. (2012). CodY orchestrates the expression of virulence determinants in emetic *Bacillus cereus* by impacting key regulatory circuits. Mol Microbiol 85, 67-88. http://doi:10.1111/j.1365-2958.2012.08090.x

Fujita, M., Gonzalez-Pastor, J.E., and Losick, R. (2005). High- and low-threshold genes in the Spo0A regulon of *Bacillus subtilis*. J Bacteriol 187, 1357-1368 http://dx.doi.org/1310.1128/ JB.1187.1354.1357-1368.2005

Fujita, Y. (2009). Carbon catabolite control of the metabolic network in *Bacillus subtilis*. Biosci Biotechnol Biochem 73, 245-259. http://doi:210.1271/bbb 80479

Fux, C.A., Costerton, J.W., Stewart, P.S., and Stoodley, P. (2005). Survival strategies of infectious biofilms. Trends Microbiol 13, 34-40. http://dx.doi.org/10.1016/j.tim.2004.1011.1010

Gallego del Sol, F., and Marina, A. (2013). Structural basis of Rap phosphatase inhibition by Phr peptides. PLoS Biol 11, e1001511. http://doi:1001510.1001371/journal.pbio.1001511

Gao, T., Foulston, L., Chai, Y., Wang, Q., and Losick, R. (2015). Alternative modes of biofilm formation by plant-associated *Bacillus cereus*. Microbiologyopen 4, 452-464. http://doi:410.1002/mbo1003.1251

Gao, X., Mukherjee, S., Matthews, P.M., Hammad, L.A., Kearns, D.B., and Dann, C.E., 3rd (2013). Functional characterization of core components of the *Bacillus subtilis* cyclic-di-GMP signaling pathway. J Bacteriol 195, 4782-4792. http://doi:4710.1128/AEM.01638-01614

Garrett, T.R., Bhakoo, M., and Zhang, Z.B. (2008). Bacterial adhesion and biofilms on surfaces. Prog Nat Sci 18, 1049-1056. http://dx.doi.org/1010.1016/j.pnsc.2008.1004.1001

Geiger, T., and Wolz, C. (2014). Intersection of the stringent response and the CodY regulon in low GC Gram-positive bacteria. Int J Med Microbiol 304, 150-155. http:// doi: 10.1016/j.ijmm.2013.11.013

Geiger, T., Goerke, C., Mainiero, M., Kraus, D., and Wolz, C. (2008). The virulence regulator *sae* of *Staphylococcus aureus*: Promoter activities and response to phagocytosis-related signals. Journal of Bacteriology 190, 3419-3428. https://doi.org/3410.1128/JB.01927-01907

Gelis-Jeanvoine, S., Canette, A., Gohar, M., Caradec, T., Lemy, C., Gominet, M., Jacques, P., Lereclus, D., and Slamti, L. (2017). Genetic and functional analyses of *krs*, a locus encoding kurstakin, a lipopeptide produced by *Bacillus thuringiensis*. Res Microbiol 168, 356-368. http://dx.doi.org/310.1016/j.resmic.2016.1006.1002

Geoghegan, J.A., Corrigan, R.M., Gruszka, D.T., Speziale, P., O'Gara, J.P., Potts, J.R., and Foster, T.J. (2010). Role of Surface Protein SasG in Biofilm Formation by *Staphylococcus aureus*. J Bacteriol 192, 5663-5673. https://doi: 10.1128/JB.00628-10

Gilbert, P., Das, J., and Foley, I. (1997). Biofilm susceptibility to antimicrobials. Adv Dent Res 11, 160-167. http://dx.doi.org/110.1177/08959374970110010701

Gohar, M., Faegri, K., Perchat, S., Ravnum, S., Okstad, O.A., Gominet, M., Kolsto, A.B., and Lereclus, D. (2008). The PlcR virulence regulon of *Bacillus cereus*. PLoS One 3, e2793. http://dx.doi.org/2710.1371/journal.pone.0002793

Gominet, M., Slamti, L., Gilois, N., Rose, M., and Lereclus, D. (2001). Oligopeptide permease is required for expression of the *Bacillus thuringiensis plcR* regulon and for virulence. Mol Microbiol 40, 963-975. http://dx.doi.org/910.1046/j.1365-2958.2001.02440.x

Gonzalez-Pastor, J.E., Hobbs, E.C., and Losick, R. (2003). Cannibalism by sporulating bacteria. Science 301, 510-513. http://dx.doi.org/510.1126/science.1086462

Grau, R.R., de Ona, P., Kunert, M., Lenini, C., Gallegos-Monterrosa, R., Mhatre, E., Vileta, D., Donato, V., Holscher, T., Boland, W., et al. (2015). A Duo of Potassium-Responsive Histidine Kinases Govern the Multicellular Destiny of *Bacillus subtilis*. MBio 6, e00581. http://dx.doi.org/00510.01128/mBio.00581-00515

Guttenplan, S.B., Blair, K.M., and Kearns, D.B. (2010). The EpsE flagellar clutch is bifunctional and synergizes with EPS biosynthesis to promote *Bacillus subtilis* biofilm formation. PLoS Genet 6, e1001243. http://dx.doi.org/1001210.1001371/journal.pgen.1001243

Hall-Stoodley, L., Costerton, J.W., and Stoodley, P. (2004). Bacterial biofilms: from the natural environment to infectious diseases. Nat Rev Microbiol 2, 95-108. http://dx.doi.org/ 110.1038/nrmicro1821

Hamon, M.A., and Lazazzera, B.A. (2001). The sporulation transcription factor Spo0A is required for biofilm development in *Bacillus subtilis*. Mol Microbiol 42, 1199-1209. http://dx.doi.org/1110.1046/j.1365-2958.2001.02709.x

Hamon, M.A., Stanley, N.R., Britton, R.A., Grossman, A.D., and Lazazzera, B.A. (2004). Identification of AbrB-regulated genes involved in biofilm formation by *Bacillus subtilis*. Mol Microbiol 52, 847-860. http://doi810.1111/j.1365-2958.2004.04023.x

Hardie, K.R., and Heurlier, K. (2008). Establishing bacterial communities by 'word of mouth': LuxS and autoinducer 2 in biofilm development. Nat Rev Microbiol 6, 635-643. http://dx.doi.org/610.1038/nrmicro1916

Hayrapetyan, H., Tempelaars, M., Nierop Groot, M., and Abee, T. (2015). *Bacillus cereus* ATCC 14579 RpoN (Sigma 54) Is a Pleiotropic Regulator of Growth, Carbohydrate Metabolism, Motility, Biofilm Formation and Toxin Production. PLoS One 10, e0134872. https://doi.org/ 0134810.0131371/journal.pone.0134872

Hobley, L., Kim, S.H., Maezato, Y., Wyllie, S., Fairlamb, A.H., Stanley-Wall, N.R., and Michael, A.J. (2014). Norspermidine is not a self-produced trigger for biofilm disassembly. Cell 156, 844-854. http://dx.doi.org/810.1016/j.cell.2014.1001.1012

Hobley, L., Ostrowski, A., Rao, F.V., Bromley, K.M., Porter, M., Prescott, A.R., MacPhee, C.E., Van Aalten, D.M.F., and Stanley-Wall, N.R. (2013). BslA is a self-assembling bacterial hydrophobin that coats the *Bacillus subtilis* biofilm. P Natl Acad Sci USA 110, 13600-13605. http://doi:13610.11073/pnas.1306390110

Hochbaum, A.I., Kolodkin-Gal, I., Foulston, L., Kolter, R., Aizenberg, J., and Losick, R. (2011). Inhibitory Effects of D-Amino Acids on *Staphylococcus aureus* Biofilm Development. Journal of Bacteriology 193, 5616-5622. http://doi:5610.1128/JB.05534-05511

Houry, A., Briandet, R., Aymerich, S., and Gohar, M. (2010). Involvement of motility and flagella in *Bacillus cereus* biofilm formation. Microbiol-Sgm 156, 1009-1018. http://dx.doi.org/1010.1099/ mic.1000.034827-034820

Hsu, R.L., Lee, K.T., Wang, J.H., Lee, L.Y.L., and Chen, R.P.Y. (2009). Amyloid-Degrading Ability of Nattokinase from *Bacillus subtilis* Natto. J Agr Food Chem 57, 503-508. http://doi:510.1021/jf803072r

Hsueh, Y.H., Somers, E.B., and Wong, A.C. (2008). Characterization of the *codY* gene and its influence on biofilm formation in *Bacillus cereus*. Arch Microbiol 189, 557-568. http://doi:510.1007/s00203-00008-00348-00208

Hsueh, Y.H., Somers, E.B., Lereclus, D., and Wong, A.C. (2006). Biofilm formation by *Bacillus cereus* is influenced by PlcR, a pleiotropic regulator. Appl Environ Microbiol 72, 5089-5092. http://doi:5010.1128/AEM.00573-00506

Huillet, E., Tempelaars, M.H., Andre-Leroux, G., Wanapaisan, P., Bridoux, L., Makhzami, S., Panbangred, W., Martin-Verstraete, I., Abee, T., and Lereclus, D. (2012). PlcRa, a new quorum-sensing regulator from *Bacillus cereus*, plays a role in oxidative stress responses and cysteine metabolism in stationary phase. PLoS One 7, e51047. http://dx.doi.org/51010.51371/ journal.pone.0051047

Hurlow, J., Couch, K., Laforet, K., Bolton, L., Metcalf, D., and Bowler, P. (2015). Clinical Biofilms: A Challenging Frontier in Wound Care. Adv Wound Care (New Rochelle) 4, 295-301. http://dx.doi.org/210.1089/wound.2014.0567

Irie, Y., O'Toole G, A., and Yuk, M.H. (2005). *Pseudomonas aeruginosa* rhamnolipids disperse *Bordetella bronchiseptica* biofilms. FEMS Microbiol Lett 250, 237-243. https://doi.org/210.1016/j.femsle.2005.1007.1012

Irigul-Sonmez, O., Koroglu, T.E., Ozturk, B., Kovacs, A.T., Kuipers, O.P., and Yazgan-Karatas, A. (2014). In *Bacillus subtilis* LutR is part of the global complex regulatory network governing the adaptation to the transition from exponential growth to stationary phase. Microbiol-Sgm 160, 243-260. https://doi:10.1099/mic.0.064675-0

Irnov, I., and Winkler, W.C. (2010). A regulatory RNA required for antitermination of biofilm and capsular polysaccharide operons in Bacillales. Mol Microbiol 76, 559-575. http://dx.doi.org/510.1111/j.1365-2958.2010.07131.x

Ivanova, N., Sorokin, A., Anderson, I., Galleron, N., Candelon, B., Kapatral, V., Bhattacharyya, A., Reznik, G., Mikhailova, N., Lapidus, A., et al. (2003). Genome sequence of *Bacillus cereus* and comparative analysis with *Bacillus anthracis*. Nature 423, 87-91. http://dx.doi.org/10.1038/nature01582

Iwase, T., Uehara, Y., Shinji, H., Tajima, A., Seo, H., Takada, K., Agata, T., and Mizunoe, Y. (2010). *Staphylococcus epidermidis* Esp inhibits *Staphylococcus aureus* biofilm formation and nasal colonization. Nature 465, 346-349. http://doi:310.1038/nature09074

Izano, E.A., Amarante, M.A., Kher, W.B., and Kaplan, J.B. (2008a). Differential roles of poly-N-acetylglucosamine surface polysaccharide and extracellular DNA in *Staphylococcus aureus* and *Staphylococcus epidermidis* biofilms. Appl Environ Microbiol 74, 470-476. http://doi:410.1128/AEM.02073-02007

Izano, E.A., Sadovskaya, I., Wang, H., Vinogradov, E., Ragunath, C., Ramasubbu, N., Jabbouri, S., Perry, M.B., and Kaplan, J.B. (2008b). Poly-N-acetylglucosamine mediates biofilm formation and detergent resistance in *Aggregatibacter actinomycetemcomitans*. Microb Pathog 44, 52-60. http://doi:10.1016/j.micpath.2007.1008.1004

Jefferson, K.K., Cramton, S.E., Gotz, F., and Pier, G.B. (2003). Identification of a 5-nucleotide sequence that controls expression of the *ica* locus in *Staphylococcus aureus* and characterization of the DNA-binding properties of IcaR. Mol Microbiol 48, 889-899. https://doi.org/10.1046/j.1365-2958.2003.03482.x

Jefferson, K.K., Pier, D.B., Goldmann, D.A., and Pier, G.B. (2004). The teicoplanin-associated locus regulator (TcaR) and the intercellular adhesin locus regulator (IcaR) are transcriptional inhibitors of the *ica* locus in *Staphylococcus aureus*. J Bacteriol 186, 2449-2456. https://doi:10.1128/JB.186.8.2449-2456.2004

Jeong, D.W., Cho, H., Lee, H., Li, C., Garza, J., Fried, M., and Bae, T. (2011). Identification of the P3 promoter and distinct roles of the two promoters of the SaeRS two-component system in *Staphylococcus aureus*. J Bacteriol 193, 4672-4684. https://doi.org/4610.1128/JB.00353-00311

Jiang, M., Shao, W., Perego, M., and Hoch, J.A. (2000). Multiple histidine kinases regulate entry into stationary phase and sporulation in *Bacillus subtilis*. Mol Microbiol 38, 535-542. http://dx.doi.org/510.1046/j. 1365-2958.2000.02148.x

Jiang, Q., Jin, Z., and Sun, B. (2018). MgrA negatively Regulates Biofilm Formation and Detachment by Repressing the Expression of *psm* Operons in *Staphylococcus aureus*. Appl Environ Microbiol 84, e01008-18. http:// doi: 10.1128/AEM.01008-18

Jones, M.B., and Blaser, M.J. (2003). Detection of a *luxS*-signaling molecule in *Bacillus anthracis*. Infect Immun 71, 3914-3919. http://dx.doi.org/3910.1128/ IAI.3971.3917.3914-3919.2003

Jones, M.B., Jani, R., Ren, D., Wood, T.K., and Blaser, M.J. (2005). Inhibition of *Bacillus anthracis* growth and virulence-gene expression by inhibitors of quorum-sensing. J Infect Dis 191, 1881-1888. http://doi:1810.1086/429696

Joo, H.S., Chatterjee, S.S., Villaruz, A.E., Dickey, S.W., Tan, V.Y., Chen, Y., Sturdevant, D.E., Ricklefs, S.M., and Otto, M. (2016). Mechanism of Gene Regulation by a *Staphylococcus aureus* Toxin. Mbio 7, e01579-16. https://doi: 10.1128/mBio.01579-16

Kaplan, J.B. (2010). Biofilm dispersal: mechanisms, clinical implications, and potential therapeutic uses. J Dent Res 89, 205-218. http://doi: 210.1177/0022034509359403

Kaplan, J.B., Meyenhofer, M.F., and Fine, D.H. (2003). Biofilm growth and detachment of *Actinobacillus actinomycetemcomitans*. J Bacteriol 185, 1399-1404. http://dx.doi.org/1310.1128/ JB.1185.1394.1399-1404.2003

Kaplan, J.B., Velliyagounder, K., Ragunath, C., Rohde, H., Mack, D., Knobloch, J.K., and Ramasubbu, N. (2004). Genes involved in the synthesis and degradation of matrix polysaccharide in *Actinobacillus actinomycetemcomitans* and *Actinobacillus pleuropneumoniae* biofilms. J Bacteriol 186, 8213-8220. http://doi:8210.1128/JB.8186.8224.8213-8220.2004

Karatan, E., and Watnick, P. (2009). Signals, regulatory networks, and materials that build and break bacterial biofilms. Microbiol Mol Biol Rev 73, 310-347. http://dx.doi.org/310.1128/MMBR.00041-00008

Karunakaran, E., and Biggs, C.A. (2011). Mechanisms of *Bacillus cereus* biofilm formation: an investigation of the physicochemical characteristics of cell surfaces and extracellular proteins. Appl Microbiol Biotechnol 89, 1161-1175. http://doi:1110.1007/s00253-00010-02919-00252

Kavanaugh, J.S., Thoendel, M., and Horswill, A.R. (2007). A role for type I signal peptidase in *Staphylococcus aureus* quorum sensing. Mol Microbiol 65, 780-798. http://dx.doi.org/710.1111/ j.1365-2958.2007.05830.x

Kearns, D.B., and Losick, R. (2003). Swarming motility in undomesticated *Bacillus subtilis*. Mol Microbiol 49, 581-590. http://dx.doi.org/510.1046/j. 1365-2958.2003.03584.x

Kearns, D.B., Chu, F., Branda, S.S., Kolter, R., and Losick, R. (2005). A master regulator for biofilm formation by *Bacillus subtilis*. Mol Microbiol 55, 739-749. http://dx.doi.org/710.1046/j.1365-2958.2003.03584.x

Kiedrowski, M.R., Crosby, H.A., Hernandez, F.J., Malone, C.L., McNamara, J.O., 2nd, and Horswill, A.R. (2014). *Staphylococcus aureus* Nuc2 is a functional, surface-attached extracellular nuclease. PLoS One 9, e95574. http://doi: 95510.91371/journal.pone.0095574

Kiedrowski, M.R., Kavanaugh, J.S., Malone, C.L., Mootz, J.M., Voyich, J.M., Smeltzer, M.S., Bayles, K.W., and Horswill, A.R. (2011). Nuclease modulates biofilm formation in community-associated methicillin-resistant *Staphylococcus aureus*. PLoS One 6, e26714. http://doi:26710.21371/journal.pone.0026714

Kim, S.K., and Lee, J.H. (2016). Biofilm dispersion in *Pseudomonas aeruginosa*. J Microbiol 54, 71-85. http://doi:10.1007/s12275-12016-15528-12277

Kinsinger, R.F., Kearns, D.B., Hale, M., and Fall, R. (2005). Genetic requirements for potassium ion-dependent colony spreading in *Bacillus subtilis*. J Bacteriol 187, 8462-8469. http://dx.doi.org/8410.1128/JB. 8187.8424.8462-8469.2005

Kinsinger, R.F., Shirk, M.C., and Fall, R. (2003). Rapid surface motility in *Bacillus subtilis* is dependent on extracellular surfactin and potassium ion. J Bacteriol 185, 5627-5631. http://dx.doi.org/5610.1128/JB. 5185.5618.5627-5631.2003

Kobayashi, K. (2007a). *Bacillus subtilis* pellicle formation proceeds through genetically defined morphological changes. J Bacteriol 189, 4920-4931. http:// dx.doi.org/4910.1128/JB.00157-00107

Kobayashi, K. (2007b). Gradual activation of the response regulator DegU controls serial expression of genes for flagellum formation and biofilm formation in *Bacillus subtilis*. Mol Microbiol 66, 395-409. http://doi:310.1111/j.1365-2958.2007.05923.x

Kobayashi, K., and Iwano, M. (2012). BslA (YuaB) forms a hydrophobic layer on the surface of *Bacillus subtilis* biofilms. Mol Microbiol 85, 51-66. http://dx.doi.org/10.1111/j.1365-2958.2012.08094.x

Koh, K.S., Lam, K.W., Alhede, M., Queck, S.Y., Labbate, M., Kjelleberg, S., and Rice, S.A. (2007). Phenotypic diversification and adaptation of Serratia marcescens MG1 biofilm-derived morphotypes. J Bacteriol 189, 119-130. http://doi:110.1128/JB.00930-00906

Kokai-Kun, J.F., Chanturiya, T., and Mond, J.J. (2009). Lysostaphin eradicates established *Staphylococcus aureus* biofilms in jugular vein catheterized mice. J Antimicrob Chemoth 64, 94-100. https://doi:10.1093/jac/dkp145

Kolodkin-Gal, I., Elsholz, A.K., Muth, C., Girguis, P.R., Kolter, R., and Losick, R. (2013). Respiration control of multicellularity in *Bacillus subtilis* by a complex of the cytochrome chain with a membrane-embedded histidine kinase. Genes Dev 27, 887-899. http://dx.doi.org/810.1101/ gad.215244.215113

Kolodkin-Gal, I., Romero, D., Cao, S., Clardy, J., Kolter, R., and Losick, R. (2010). D-amino acids trigger biofilm disassembly. Science 328, 627-629. http://dx.doi.org/10.1126/science.1188628

Kolter, R., and Greenberg, E.P. (2006). Microbial sciences: the superficial life of microbes. Nature 441, 300-302. http://dx.doi.org/10.1038/441300a

Kostakioti, M., Hadjifrangiskou, M., and Hultgren, S.J. (2013). Bacterial biofilms: development, dispersal, and therapeutic strategies in the dawn of the postantibiotic era. Cold Spring Harb Perspect Med 3, a010306. http://dx.doi.org/10.1101/cshperspect.a010306

Landini, P., Antoniani, D., Burgess, J.G., and Nijland, R. (2010). Molecular mechanisms of compounds affecting bacterial biofilm formation and dispersal. Appl Microbiol Biot 86, 813-823. https://doi:10.1007/s00253-010-2468-8

Lanigan-Gerdes, S., Dooley, A.N., Faull, K.F., and Lazazzera, B.A. (2007). Identification of subtilisin, Epr and Vpr as enzymes that produce CSF, an extracellular signalling peptide of *Bacillus subtilis*. Mol Microbiol 65, 1321-1333. http://dx.doi.org/10.1111/j.1365-2958.2007.05869.x

Lauderdale, K.J., Malone, C.L., Boles, B.R., Morcuende, J., and Horswill, A.R. (2010). Biofilm Dispersal of Community-Associated Methicillin-Resistant *Staphylococcus aureus* on Orthopedic Implant Material. J Orthop Res 28 55-61. https://doi:10.1002/jor.20943

Lazarevic, V., Soldo, B., Medico, N., Pooley, H., Bron, S., and Karamata, D. (2005). *Bacillus subtilis* alpha-phosphoglucomutase is required for normal cell morphology and biofilm formation. Appl Environ Microbiol 71, 39-45. http://doi 10.1128/AEM.71.1.39-45.2005

Lazazzera, B.A., Kurtser, I.G., McQuade, R.S., and Grossman, A.D. (1999). An autoregulatory circuit affecting peptide signaling in *Bacillus subtilis*. J Bacteriol 181, 5193-5200.

LeDeaux, J.R., Solomon, J.M., and Grossman, A.D. (1997). Analysis of non-polar deletion mutations in the genes of the *spo0K* (*opp*) operon of *Bacillus subtilis*. FEMS Microbiol Lett 153, 63-69. http://dx.doi.org/10.1111/j. 1574-6968.1997.tb10464.x

Lee, K., Costerton, J.W., Rave, J., Auerbach, R.K., Wagner, D.M., Keim, P., and Leid, J.G. (2007). Phenotypic and functional characterization of *Bacillus anthracis* biofilms. Microbiol-Sgm 153, 1693-1701. http://doi:10.1099/mic. 0.2006/003376-0

Leiman, S.A., May, J.M., Lebar, M.D., Kahne, D., Kolter, R., and Losick, R. (2013). D-amino acids indirectly inhibit biofilm formation in *Bacillus subtilis* by interfering with protein synthesis. J Bacteriol 195, 5391-5395. http://doi: 10.1128/JB.00975-13

Lemon, K.P., Earl, A.M., Vlamakis, H.C., Aguilar, C., and Kolter, R. (2008). Biofilm development with an emphasis on *Bacillus subtilis*. Curr Top Microbiol Immunol 322, 1-16. http://dx.doi.org/10.1007/978-3-540-75418-3_1

Li, M., Cha, D.J., Lai, Y.P., Villaruz, A.E., Sturdevant, D.E., and Otto, M. (2007). The antimicrobial peptide-sensing system aps of *Staphylococcus aureus*. Mol Microbiol 66, 1136-1147. https://doi.org/10.1111/j.1365-2958.2007.05986.x

Lindback, T., Mols, M., Basset, C., Granum, P.E., Kuipers, O.P., and Kovacs, A.T. (2012). CodY, a pleiotropic regulator, influences multicellular behaviour and efficient production of virulence factors in *Bacillus cereus*. Environ Microbiol 14, 2233-2246. http://doi:10.1111/j.1462-2920.2012.02766.x

Lister, J.L., and Horswill, A.R. (2014). *Staphylococcus aureus* biofilms: recent developments in biofilm dispersal. Front Cell Infect Microbiol 4, 178. http://doi: 10.3389/fcimb.2014.00178

Liu, Q., Wang, X., Qin, J., Cheng, S., Yeo, W.S., He, L., Ma, X., Liu, X., Li, M., and Bae, T. (2017). The ATP-Dependent Protease ClpP Inhibits Biofilm Formation by Regulating Agr and Cell Wall Hydrolase Sle1 in *Staphylococcus aureus*. Front Cell Infect Microbiol 7, 181. https://doi.org/10.3389/fcimb.2017.00181

Lombardia, E., Rovetto, A.J., Arabolaza, A.L., and Grau, R.R. (2006). A LuxS-dependent cell-to-cell language regulates social behavior and development in *Bacillus subtilis*. J Bacteriol 188, 4442-4452. http://dx.doi.org/10.1128/JB.00165-06

Lopez, D., and Kolter, R. (2010). Extracellular signals that define distinct and coexisting cell fates in *Bacillus subtilis*. FEMS Microbiol. Rev 34, 134-149. http://doi:10.1111/j.1574-6976.2009.00199.x

Lopez, D., Fischbach, M.A., Chu, F., Losick, R., and Kolter, R. (2009a). Structurally diverse natural products that cause potassium leakage trigger multicellularity in *Bacillus subtilis*. Proc Natl Acad Sci USA 106, 280-285. http://doi: 10.1073/pnas.0810940106

Lopez, D., Vlamakis, H., Losick, R., and Kolter, R. (2009b). Paracrine signaling in a bacterium. Genes Dev 23, 1631-1638. http://doi:10.1101/gad.1813709

Lopez, D., Vlamakis, H., Losick, R., and Kolter, R. (2009c). Cannibalism enhances biofilm development in *Bacillus subtilis*. Mol Microbiol 74, 609-618. http://doi:10.1111/j.1365-2958.2009.06882.x

Lorca, G.L., Chung, Y.J., Barabote, R.D., Weyler, W., Schilling, C.H., and Saier, M.H., Jr. (2005). Catabolite repression and activation in *Bacillus subtilis*: dependency on CcpA, HPr, and HprK. J Bacteriol 187, 7826-7839. http://doi: 10.1128/JB.187.22.7826-7839.2005

Ma, L., Conover, M., Lu, H., Parsek, M.R., Bayles, K., and Wozniak, D.J. (2009). Assembly and development of the *Pseudomonas aeruginosa* biofilm matrix. PLoS Pathog 5, e1000354. http://doi:10.1371/journal.ppat.1000354

Magnuson, R., Solomon, J., and Grossman, A.D. (1994). Biochemical and genetic characterization of a competence pheromone from *B. subtilis*. Cell 77, 207-216.

Mainiero, M., Goerke, C., Geiger, T., Gonser, C., Herbert, S., and Wolz, C. (2010). Differential target gene activation by the *Staphylococcus aureus* two-component system *saeRS*. J Bacteriol 192, 613-623. https://doi.org/10 .1128/ JB.01242-09

Majed, R., Faille, C., Kallassy, M., and Gohar, M. (2016). *Bacillus cereus* Biofilms-Same, Only Different. Front Microbiol 7, 1054. https://doi.org/ 10.1016/0006-291X(67)90090-3

Makgotlho, P.E., Marincola, G., Schafer, D., Liu, Q., Bae, T., Geiger, T., Wasserman, E., Wolz, C., Ziebuhr, W., and Sinha, B. (2013). SDS interferes with SaeS signaling of *Staphylococcus aureus* independently of SaePQ. PLoS One 8, e71644. http://dx.doi.org/10.1371/journal.pone.0071644

Mann, E.E., Rice, K.C., Boles, B.R., Endres, J.L., Ranjit, D., Chandramohan, L., Tsang, L.H., Smeltzer, M.S., Horswill, A.R., and Bayles, K.W. (2009). Modulation of eDNA release and degradation affects *Staphylococcus aureus* biofilm maturation. PLoS One 4, e5822. http://doi:10.1371/journal. pone. 0005822

Mashruwala, A.A., Gries, C.M., Scherr, T.D., Kielian, T., and Boyd, J.M. (2017a). SaeRS Is Responsive to Cellular Respiratory Status and Regulates Fermentative Biofilm Formation in *Staphylococcus aureus*. Infect Immun 85, e00157. http://doi:10.1128/IAI.00157-17

Mashruwala, A.A., Guchte, A.V., and Boyd, J.M. (2017b). Impaired respiration elicits SrrAB-dependent programmed cell lysis and biofilm formation in *Staphylococcus aureus*. Elife 6, e23845. http://doi:10.7554/eLife.23845

McDougald, D., Rice, S.A., Barraud, N., Steinberg, P.D., and Kjelleberg, S. (2011). Should we stay or should we go: mechanisms and ecological consequences for biofilm dispersal. Nat Rev Microbiol 10, 39-50. http:// dx.doi.org/10.1038/nrmicro2695

McLoon, A.L., Kolodkin-Gal, I., Rubinstein, S.M., Kolter, R., and Losick, R. (2011). Spatial regulation of histidine kinases governing biofilm formation in *Bacillus subtilis*. J Bacteriol 193, 679-685. http://doi:10.1128/JB.01186-10

Merino, N., Toledo-Arana, A., Vergara-Irigaray, M., Valle, J., Solano, C., Calvo, E., Lopez, J.A., Foster, T.J., Penades, J.R., and Lasa, I. (2009). Protein A-Mediated Multicellular Behavior in *Staphylococcus aureus*. J Bacteriol 191, 832-843. https://doi:10.1128/JB.01222-08

Meyer, F.M., Jules, M., Mehne, F.M.P., Le Coq, D., Landmann, J.J., Gorke, B., Aymerich, S., and Stulke, J. (2011). Malate-Mediated Carbon Catabolite Repression in *Bacillus subtilis* Involves the HPrK/CcpA Pathway. J Bacteriol 193, 6939-6949. https://doi:10.1128/Jb.06197-11

Mhatre, E., Monterrosa, R.G., and Kovacs, A.T. (2014). From environmental signals to regulators: modulation of biofilm development in Gram-positive bacteria. J Basic Microbiol 54, 616-632. http://dx.doi.org/10.1002/jobm.201400175

Mhatre, E., Sundaram, A., Holscher, T., Muhlstadt, M., Bossert, J., and Kovacs, A.T. (2017). Presence of Calcium Lowers the Expansion of *Bacillus subtilis* Colony Biofilms. Microorganisms 5. http://dx.doi.org/10.3390/microorganisms5010007

Mielich-Suss, B., Schneider, J., and Lopez, D. (2013). Overproduction of flotillin influences cell differentiation and shape in *Bacillus subtilis*. MBio 4, e00719-00713. http://dx.doi.org/10.1128/mBio.00719-13

Miller, M.B., and Bassler, B.L. (2001). Quorum sensing in bacteria. Annu Rev Microbiol 55, 165-199. http://dx.doi.org/10.1146/annurev.micro.55.1.165

Mirani, Z.A., Aziz, M., Khan, M.N., Lal, I., ul Hassan, N., and Khan, S.I. (2013). Biofilm formation and dispersal of *Staphylococcus aureus* under the influence of oxacillin. Microb Pathogenesis 61-62, 66-72. https://doi.org/10.1016/j.micpath.2013.05.002

Mirani, Z.A., Khan, M.N., Siddiqui, A., Khan, F., Aziz, M., Naz, S., Ahmed, A., and Khan, S.I. (2018). Ascorbic acid augments colony spreading by reducing biofilm formation of methicillin-resistant *Staphylococcus aureus*. Iran J Basic Med Sci 21, 175-180. https://doi:10.22038/IJBMS.2018.20714.5398

Mironczuk, A.M., Kovacs, A.T., and Kuipers, O.P. (2008). Induction of natural competence in *Bacillus cereus* ATCC14579. Microb Biotechnol 1, 226-235. http://doi:10.1111/j.1751-7915.2008.00023.x

Mitchell, G., Fugere, A., Gaudreau, K.P., Brouillette, E., Frost, E.H., Cantin, A.M., and Malouin, F. (2013). SigB Is a Dominant Regulator of Virulence in *Staphylococcus aureus* Small-Colony Variants. Plos One 8, e65018. http:// doi:10.1371/journal.pone.0065018

Molle, V., Fujita, M., Jensen, S.T., Eichenberger, P., Gonzalez-Pastor, J.E., Liu, J.S., and Losick, R. (2003). The Spo0A regulon of *Bacillus subtilis*. Mol Microbiol 50, 1683-1701. https://doi.org/10.1046/j.1365-2958.2003.03818.x

Mootz, J.M., Malone, C.L., Shaw, L.N., and Horswill, A.R. (2013). Staphopains Modulate *Staphylococcus aureus* Biofilm Integrity. Infect Immun 81, 3227-3238. https://doi:10.1128/IAI.00377-13

Morikawa, M., Kagihiro, S., Haruki, M., Takano, K., Branda, S., Kolter, R., and Kanaya, S. (2006). Biofilm formation by a *Bacillus subtilis* strain that produces gamma-polyglutamate. Microbiology 152, 2801-2807. http://doi: 10.1099/mic. 0.29060-0

Novick, R.P., and Geisinger, E. (2008). Quorum sensing in *staphylococci*. Annu Rev Genet 42, 541-564. http://doi:10.1146/annurev.genet.42.110807.091640

Novick, R.P., and Jiang, D. (2003). The *staphylococcal* saeRS system coordinates environmental signals with *agr* quorum sensing. Microbiology 149, 2709-2717. http://dx.doi.org/10.1099/ mic.0.26575-0

Ogura, M., and Tsukahara, K. (2010). Autoregulation of the *Bacillus subtilis* response regulator gene *degU* is coupled with the proteolysis of DegU-P by ClpCP. Mol Microbiol 75, 1244-1259. http://doi:10.1128/JB.01881-12

Oknin, H., Steinberg, D., and Shemesh, M. (2015). Magnesium ions mitigate biofilm formation of *Bacillus* species via downregulation of matrix genes expression. Front Microbiol 6, 907. http://dx.doi.org/10.3389/fmicb.2015.00907

Okshevsky, M., and Meyer, R.L. (2015). The role of extracellular DNA in the establishment, maintenance and perpetuation of bacterial biofilms. Crit Rev Microbiol 41, 341-352. https://doi:10.3109/1040841X.2013.841639

Olson, M.E., Nygaard, T.K., Ackermann, L., Watkins, R.L., Zurek, O.W., Pallister, K.B., Griffith, S., Kiedrowski, M.R., Flack, C.E., Kavanaugh, J.S., et al. (2013). *Staphylococcus aureus* nuclease is an SaeRS-dependent virulence factor. Infect Immun 81, 1316-1324. http://doi: 10.1128/IAI.01242-12

Omer Bendori, S., Pollak, S., Hizi, D., and Eldar, A. (2015). The RapP-PhrP quorum-sensing system of *Bacillus subtilis* strain NCIB3610 affects biofilm formation through multiple targets, due to an atypical signal-insensitive allele of RapP. J Bacteriol 197, 592-602. http://dx.doi.org/ 10.1128/JB.02382-14

O'Neill, E., Pozzi, C., Houston, P., Humphreys, H., Robinson, D.A., Loughman, A., Foster, T.J., and O'Gara, J.P. (2008). A novel *Staphylococcus aureus* biofilm phenotype mediated by the fibronectin-binding proteins, FnBPA and FnBPB. J Bacteriol 190, 3835-3850. https://doi:10.1128/JB.00167-08

O'Neill, E., Pozzi, C., Houston, P., Smyth, D., Humphreys, H., Robinson, D.A., and O'Gara, J.P. (2007). Association between methicillin susceptibility and biofilm regulation in *Staphylococcus aureus* isolates from device-related infections. J Clin Microbiol 45, 1379-1388. https://doi:10.1128/JCM.02280-06

O'Toole, G.A., and Kolter, R. (1998). Initiation of biofilm formation in *Pseudomonas fluorescens* WCS365 proceeds via multiple, convergent signalling pathways: a genetic analysis. Mol Microbiol 28, 449-461. http://dx.doi.org/10.1046/j.1365-2958.1998.00797.x

Pamp, S.J., and Tolker-Nielsen, T. (2007). Multiple roles of biosurfactants in structural biofilm development by *Pseudomonas aeruginosa*. J Bacteriol 189, 2531-2539. https://doi:10.1128/JB.01515-06

Pamp, S.J., Frees, D., Engelmann, S., Hecker, M., and Ingmer, H. (2006). Spx is a global effector impacting stress tolerance and biofilm formation in *Staphylococcus aureus*. J Bacteriol 188, 4861-4870. https://doi:10.1128/JB.00194-06

Pane-Farre, J., Jonas, B., Forstner, K., Engelmann, S., and Hecker, M. (2006). The sigma(B) regulon in *Staphylococcus aureus* and its regulation. Int J Med Microbiol 296, 237-258. http://10.1016/ j.ijmm.2005.11.011

Parashar, V., Konkol, M.A., Kearns, D.B., and Neiditch, M.B. (2013). A plasmid-encoded phosphatase regulates *Bacillus subtilis* biofilm architecture, sporulation, and genetic competence. J Bacteriol 195, 2437-2448. http://dx.doi.org/10.1128/JB.02030-12

Parashar, V., Mirouze, N., Dubnau, D.A., and Neiditch, M.B. (2011). Structural basis of response regulator dephosphorylation by Rap phosphatases. PLoS Biol 9, e1000589. http://dx.doi.org/ 10.1371/journal.pbio.1000589

Parsek, M.R., and Singh, P.K. (2003). Bacterial biofilms: an emerging link to disease pathogenesis. Annu Rev Microbiol 57, 677-701. http://dx.doi.org/10.1146/annurev.micro.57.030502.090720

Pasvolsky, R., Zakin, V., Ostrova, I., and Shemesh, M. (2014). Butyric acid released during milk lipolysis triggers biofilm formation of *Bacillus* species. Int J Food Microbiol 181, 19-27. http://dx.doi.org/10.1016/j.ijfoodmicro.2014.04.013

Perchat, S., Dubois, T., Zouhir, S., Gominet, M., Poncet, S., Lemy, C., Aumont-Nicaise, M., Deutscher, J., Gohar, M., Nessler, S., et al. (2011). A cell-cell communication system regulates protease production during sporulation in bacteria of the *Bacillus cereus* group. Mol Microbiol 82, 619-633. http://dx.doi.org/10.1111/j.1365-2958.2011.07839.x

Perego, M. (2013). Forty years in the making: understanding the molecular mechanism of peptide regulation in bacterial development. PLoS Biol 11, e1001516. http://dx.doi.org/10.1128/ JB.185.16.4861-4871.2003

Periasamy, S., Joo, H.S., Duong, A.C., Bach, T.H.L., Tan, V.Y., Chatterjee, S.S., Cheung, G.Y.C., and Otto, M. (2012). How *Staphylococcus aureus* biofilms develop their characteristic structure. P Natl Acad Sci USA 109, 1281-1286. https://doi:10.1073/pnas.1115006109

Pflughoeft, K.J., Sumby, P., and Koehler, T.M. (2011). *Bacillus anthracis sin* locus and regulation of secreted proteases. J Bacteriol 193, 631-639. http://dx.doi.org/10.1128/JB.01083-10

Pihl, M., Arvidsson, A., Skepo, M., Nilsson, M., Givskov, M., Tolker-Nielsen, T., Svensater, G., and Davies, J.R. (2013). Biofilm formation by *Staphylococcus epidermidis* on peritoneal dialysis catheters and the effects of extracellular products from *Pseudomonas aeruginosa*. Pathog Dis 67, 192-198. http://doi:10.1111/2049-632X.12035

Pottathil, M., and Lazazzera, B.A. (2003). The extracellular Phr peptide-Rap phosphatase signaling circuit of *Bacillus subtilis*. Front Biosci 8, 32-45.

Poupel, O., Proux, C., Jagla, B., Msadek, T., and Dubrac, S. (2018). SpdC, a novel virulence factor, controls histidine kinase activity in *Staphylococcus aureus*. Plos Pathog 14, e1006917. https://doi:10.1371/journal.ppat.1006917

Queck, S.Y., Jameson-Lee, M., Villaruz, A.E., Bach, T.H.L., Khan, B.A., Sturdevant, D.E., Ricklefs, S.M., Li, M., and Otto, M. (2008). RNAIII-Independent Target Gene Control by the *agr* Quorum-Sensing System: Insight into the Evolution of Virulence Regulation in *Staphylococcus aureus*. Mol Cell 32, 150-158. http:// doi: 10.1016/j.molcel.2008.08.005

Rachid, S., Ohlsen, K., Wallner, U., Hacker, J., Hecker, M., and Ziebuhr, W. (2000). Alternative transcription factor sigma(B) is involved in regulation of biofilm expression in a *Staphylococcus aureus* mucosal isolate. J Bacteriol 182, 6824-6826. http://dx.doi.org/10.1128/JB.182.23.6824-6826.2000

Ramarao, N., and Lereclus, D. (2005). The InhA1 metalloprotease allows spores of the *B. cereus* group to escape macrophages. Cell Microbiol 7, 1357-1364. http://dx.doi.org/10.1111/j.1462-5822.2005.00562.x

Rice, K.C., Mann, E.E., Endres, J.L., Weiss, E.C., Cassat, J.E., Smeltzer, M.S., and Bayles, K.W. (2007). The *cidA* murein hydrolase regulator contributes to DNA release and biofilm development in *Staphylococcus aureus*. Proc Natl Acad Sci USA 104, 8113-8118. http://doi:10.1073/pnas.0610226104

Romero, D., Aguilar, C., Losick, R., and Kolter, R. (2010). Amyloid fibers provide structural integrity to *Bacillus subtilis* biofilms. Proc Natl Acad Sci USA 107, 2230-2234. https://doi.org/10.1073/pnas.0910560107

Romero, D., Vlamakis, H., Losick, R., and Kolter, R. (2011). An accessory protein required for anchoring and assembly of amyloid fibres in B. subtilis biofilms. Mol Microbiol 80, 1155-1168. http://doi:10.1111/j.1365-2958.2011.07653.x

Sass, P., and Bierbaum, G. (2009). Native *graS* mutation supports the susceptibility of *Staphylococcus aureus* strain SG511 to antimicrobial peptides. Int J Med Microbiol 299, 313-322. https://doi:10.1016/j.ijmm.2008.10.005

Schauder, S., Shokat, K., Surette, M.G., and Bassler, B.L. (2001). The LuxS family of bacterial autoinducers: biosynthesis of a novel quorum-sensing signal molecule. Mol Microbiol 41, 463-476. http://dx.doi.org/10.1046/j.1365-2958.2001.02532.x

Schleif, R. (2010). AraC protein, regulation of the l-arabinose operon in *Escherichia coli*, and the light switch mechanism of AraC action. Fems Microbiol Rev 34, 779-796. https://doi:10.1111/j.1574-6976.2010.00226.x

Schroeder, K., Jularic, M., Horsburgh, S.M., Hirschhausen, N., Neumann, C., Bertling, A., Schulte, A., Foster, S., Kehrel, B.E., Peters, G., et al. (2009). Molecular Characterization of a Novel *Staphylococcus aureus* Surface Protein (SasC) Involved in Cell Aggregation and Biofilm Accumulation. Plos One 4, e7567. https://doi.org/10.1371/journal.pone.0007567

Schuch, R., and Fischetti, V.A. (2009). The Secret Life of the Anthrax Agent *Bacillus anthracis*: Bacteriophage-Mediated Ecological Adaptations. Plos One 4, e6532. https://doi.org/10.1371/journal.pone.0006532

Schwartz, K., Sekedat, M.D., Syed, A.K., O'Hara, B., Payne, D.E., Lamb, A., and Boles, B.R. (2014). The AgrD N-terminal leader peptide of *Staphylococcus aureus* has cytolytic and amyloidogenic properties. Infect Immun 82, 3837-3844. http://dx.doi.org/10.1128/IAI.02111-14

Senn, M.M., Giachino, P., Homerova, D., Steinhuber, A., Strassner, J., Kormanec, J., Fluckiger, U., Berger-Bachi, B., and Bischoff, M. (2005). Molecular analysis and organization of the sigma(B) operon in *Staphylococcus aureus*. J Bacteriol 187, 8006-8019. https:/doi: 10.1128/JB.187.23.8006-8019.2005

Sepehr, S., Rahmani-Badi, A., and Babaie-Naiej, H. (2014). Unsaturated Fatty Acid, cis-2-Decenoic Acid, in Combination with Disinfectants or Antibiotics Removes Pre-Established Biofilms Formed by Food-Related Bacteria. Plos One 9, e101677. https://doi.org/10.1371/journal.pone.0101677

Shafikhani, S.H., Mandic-Mulec, I., Strauch, M.A., Smith, I., and Leighton, T. (2002). Postexponential regulation of *sin* operon expression in *Bacillus subtilis*. J Bacteriol 184, 564-571. http://dx.doi.org/10.1128/JB.184.2.564-571.2002

Shak, S., Capon, D.J., Hellmiss, R., Marsters, S.A., and Baker, C.L. (1990). Recombinant human DNase I reduces the viscosity of cystic fibrosis sputum. Proc Natl Acad Sci USA 87, 9188-9192. https://doi.org/10.1073/pnas.87.23.9188

Shakir, S.M., Bryant, K.M., Larabee, J.L., Hamm, E.E., Lovchik, J., Lyons, C.R., and Ballard, J.D. (2010). Regulatory Interactions of a Virulence-Associated Serine/Threonine Phosphatase-Kinase Pair in *Bacillus anthracis*. J Bacteriol 192, 400-409. http://doi:10.1128/JB.01221-09

Shanks, R.M.Q., Meehl, M.A., Brothers, K.M., Martinez, R.M., Donegan, N.P., Graber, M.L., Cheung, A.L., and O'Toole, G.A. (2008). Genetic evidence for an alternative citrate-dependent biofilm formation pathway in *Staphylococcus aureus* that is dependent on fibronectin binding proteins and the GraRS two-component regulatory system. Infect Immun 76, 2469-2477. https://doi: 10.1128/IAI.01370-07

Shemesh, M., Kolter, R., and Losick, R. (2010). The biocide chlorine dioxide stimulates biofilm formation in *Bacillus subtilis* by activation of the histidine kinase KinC. J Bacteriol 192, 6352-6356. http://dx.doi.org/10.1128/JB.01025-10

Shi, X.B., Rao, N.N., and Kornberg, A. (2004). Inorganic polyphosphate in *Bacillus cereus*: Motility, biofilm formation, and sporulation. P Natl Acad Sci USA 101, 17061-17065. http://doi: 10.1073/pnas.0407787101

Slamti, L., and Lereclus, D. (2005). Specificity and polymorphism of the PlcR-PapR quorum-sensing system in the *Bacillus cereus* group. J Bacteriol 187, 1182-1187. http://dx.doi.org/10.1128/ JB.187.3.1182-1187.2005

Smits, W.K., Bongiorni, C., Veening, J.W., Hamoen, L.W., Kuipers, O.P., and Perego, M. (2007). Temporal separation of distinct differentiation pathways by a dual specificity Rap-Phr system in *Bacillus subtilis*. Mol Microbiol 65, 103-120. http://doi:10.1111/j.1365-2958.2007.05776.x

Solano, C., Echeverz, M., and Lasa, I. (2014). Biofilm dispersion and quorum sensing. Curr Opin Microbiol 18, 96-104. http://dx.doi.org/10.1016/j.mib.2014.02.008

Stacy, A., Abraham, N., Jorth, P., and Whiteley, M. (2016). Microbial Community Composition Impacts Pathogen Iron Availability during Polymicrobial Infection. PLoS Pathog 12, e1006084. http://dx.doi.org/10.1371/journal.ppat.1006084

Stanley, N.R., and Lazazzera, B.A. (2004). Environmental signals and regulatory pathways that influence biofilm formation. Mol Microbiol 52, 917-924. http://dx.doi.org/10.1111/j.1365-2958.2004.04036.x

Stanley, N.R., and Lazazzera, B.A. (2005). Defining the genetic differences between wild and domestic strains of *Bacillus subtilis* that affect poly-gamma-dl-glutamic acid production and biofilm formation. Mol Microbiol 57, 1143-1158. http://doi:10.1111/j.1365-2958.2005.04746.x

Stanley, N.R., Britton, R.A., Grossman, A.D., and Lazazzera, B.A. (2003). Identification of catabolite repression as a physiological regulator of biofilm formation by *Bacillus subtilis* by use of DNA microarrays. J Bacteriol 185, 1951-1957. http://doi:10.1128/JB.185.6.1951-1957.2003

Stephenson, S., Mueller, C., Jiang, M., and Perego, M. (2003). Molecular analysis of Phr peptide processing in *Bacillus subtilis*. J Bacteriol 185, 4861-4871. http://dx.doi.org/10.1128/ JB.185.16.4861-4871.2003

Stover, A.G., and Driks, A. (1999). Secretion, localization, and antibacterial activity of TasA, a *Bacillus subtilis* spore-associated protein. J Bacteriol 181, 1664-1672.

Strauch, M.A., Bobay, B.G., Cavanagh, J., Yao, F., Wilson, A., and Le Breton, Y. (2007). Abh and AbrB control of *Bacillus subtilis* antimicrobial gene expression. J Bacteriol 189, 7720-7732. http://doi:10.1128/JB.01081-07

Sugimoto, S., Iwamoto, T., Takada, K., Okuda, K., Tajima, A., Iwase, T., and Mizunoe, Y. (2013). *Staphylococcus epidermidis* Esp degrades specific proteins associated with *Staphylococcus aureus* biofilm formation and host-pathogen interaction. J Bacteriol 195, 1645-1655. http://doi:10.1128/JB.01672-12

Taga, M.E., and Bassler, B.L. (2003). Chemical communication among bacteria. Proc Natl Acad Sci USA 100 Suppl 2, 14549-14554. http://dx.doi.org/10.1073/pnas.1934514100

Tang, J., Zhou, R., Shi, X., Kang, M., Wang, H., and Chen, H. (2008). Two thermostable nucleases coexisted in *Staphylococcus aureus*: evidence from mutagenesis and in vitro expression. FEMS Microbiol Lett 284, 176-183. http://doi:10.1111/j.1574-6968.2008.01194.x

Terra, R., Stanley-Wall, N.R., Cao, G., and Lazazzera, B.A. (2012). Identification of *Bacillus subtilis* SipW as a bifunctional signal peptidase that controls surface-adhered biofilm formation. J Bacteriol 194, 2781-2790. http://doi:10.1128/JB.06780-11

Tjalsma, H., Stover, A.G., Driks, A., Venema, G., Bron, S., and van Dijl, J.M. (2000). Conserved serine and histidine residues are critical for activity of the ER-type signal peptidase SipW of *Bacillus subtilis*. J Biol Chem 275, 25102-25108. http://doi: 10.1074/jbc.M002676200

Tran, L.S., Nagai, T., and Itoh, Y. (2000). Divergent structure of the ComQXPA quorum-sensing components: molecular basis of strain-specific communication mechanism in *Bacillus subtilis*. Mol Microbiol 37, 1159-1171. https://doi.org/10.1046/j.1365-2958.2000.02069.x

Tran, S.L., Guillemet, E., Gohar, M., Lereclus, D., and Ramarao, N. (2010). CwpFM (EntFM) is a *Bacillus cereus* potential cell wall peptidase implicated in adhesion, biofilm formation, and virulence. J Bacteriol 192, 2638-2642. http://doi:10.1128/JB.01315-09

Trejo, M., Douarche, C., Bailleux, V., Poulard, C., Mariot, S., Regeard, C., and Raspaud, E. (2013). Elasticity and wrinkled morphology of *Bacillus subtilis* pellicles. Proc Natl Acad Sci USA 110, 2011-2016. http://dx.doi.org/10.1073/pnas.1217178110

Tsang, L.H., Cassat, J.E., Shaw, L.N., Beenken, K.E., and Smeltzer, M.S. (2008). Factors Contributing to the Biofilm-Deficient Phenotype of *Staphylococcus aureus* sarA Mutants. Plos One 3, e3361. https://doi.org/10.1371/journal.pone.0003361

Valle, J., Toledo-Arana, A., Berasain, C., Ghigo, J.M., Amorena, B., Penades, J.R., and Lasa, I. (2003). SarA and not sigma(B) is essential for biofilm development by *Staphylococcus aureus*. Mol Microbiol 48, 1075-1087. https://doi.org/10.1046/j.1365-2958.2003.03493.x

Vasudevan, R. (2014). Biofilms: Microbial Cities of Scientific Significance. J Microbiol Exp 1(3), 00014. https://doi:10.15406/jmen.2014.01.00014

Veening, J.W., Hamoen, L.W., and Kuipers, O.P. (2005). Phosphatases modulate the bistable sporulation gene expression pattern in *Bacillus subtilis*. Mol Microbiol 56, 1481-1494. http://doi: 10.1111/j.1365-2958.2005.04659.x

Verhamme, D.T., Kiley, T.B., and Stanley-Wall, N.R. (2007). DegU co-ordinates multicellular behaviour exhibited by *Bacillus subtilis*. Mol Microbiol 65, 554-568. http://doi: 10.1111/j.1365-2958.2007.05810.x

Verhamme, D.T., Murray, E.J., and Stanley-Wall, N.R. (2009). DegU and Spo0A jointly control transcription of two loci required for complex colony development by *Bacillus subtilis*. J Bacteriol 191, 100-108. http://doi: 10.1128/JB.01236-08

Verplaetse, E., Slamti, L., Gohar, M., and Lereclus, D. (2015). Cell Differentiation in a *Bacillus thuringiensis* Population during Planktonic Growth, Biofilm Formation, and Host Infection. Mbio 6, e00138-15. https://doi:10.1128/mBio.00138-15

Verstraeten, N., Braeken, K., Debkumari, B., Fauvart, M., Fransaer, J., Vermant, J., and Michiels, J. (2008). Living on a surface: swarming and biofilm formation. Trends Microbiol 16, 496-506. http://dx.doi.org/10.1016/j.tim.2008.07.004

Vilain, S., Pretorius, J.M., Theron, J., and Brozel, V.S. (2009). DNA as an Adhesin: *Bacillus cereus* Requires Extracellular DNA To Form Biofilms. Appl Environ Microb 75, 2861-2868. https://doi:10.1128/AEM.01317-08

Vlamakis, H., Aguilar, C., Losick, R., and Kolter, R. (2008). Control of cell fate by the formation of an architecturally complex bacterial community. Genes Dev 22, 945-953. http://dx.doi.org/10.1101/gad.1645008

Vlamakis, H., Chai, Y., Beauregard, P., Losick, R., and Kolter, R. (2013). Sticking together: building a biofilm the *Bacillus subtilis* way. Nat Rev Microbiol 11, 157-168. http://dx.doi.org/10.1101/gad.1645008

Webb, J.S., Thompson, L.S., James, S., Charlton, T., Tolker-Nielsen, T., Koch, B., Givskov, M., and Kjelleberg, S. (2003). Cell death in *Pseudomonas aeruginosa* biofilm development. J Bacteriol 185, 4585-4592. http://doi: 10.1128/JB.185.15.4585-4592.2003

Wijman, J.G., de Leeuw, P.P., Moezelaar, R., Zwietering, M.H., and Abee, T. (2007). Air-liquid interface biofilms of *Bacillus cereus*: formation, sporulation, and dispersion. Appl Environ Microbiol 73, 1481-1488. http://dx.doi.org/10.1128/AEM.01781-06

Wilking, J.N., Zaburdaev, V., De Volder, M., Losick, R., Brenner, M.P., and Weitz, D.A. (2013). Liquid transport facilitated by channels in *Bacillus subtilis* biofilms. Proc Natl Acad Sci USA 110, 848-852. http://dx.doi.org/10.1073/pnas.1216376110

Winkelman, J.T., Blair, K.M., and Kearns, D.B. (2009). RemA (YlzA) and RemB (YaaB) regulate extracellular matrix operon expression and biofilm formation in *Bacillus subtilis*. J Bacteriol 191, 3981-3991. http://dx.doi.org/10.1128/JB.00278-09

Yan, F., Yu, Y.Y., Gozzi, K., Chen, Y., Guo, J.H., and Chai, Y.R. (2017). Genome-Wide Investigation of Biofilm Formation in *Bacillus cereus*. Appl Environ Microb 83, e00561-17. https://doi:10.1128/AEM.00561-17

Yan, F., Yu, Y.Y., Wang, L.Y., Luo, Y.M., Guo, J.H., and Chai, Y.R. (2016). The comER Gene Plays an Important Role in Biofilm Formation and Sporulation in both *Bacillus subtilis* and *Bacillus cereus*. Front Microbiol 7, 1025. https://doi:10.3389/fmicb.2016.01025

Yarwood, J.M., Bartels, D.J., Volper, E.M., and Greenberg, E.P. (2004). Quorum sensing in *Staphylococcus aureus* biofilms. J Bacteriol 186, 1838-1850. http://dx.doi.org/10.1128/JB.186.6.1838-1850.2004

Yepes, A., Schneider, J., Mielich, B., Koch, G., Garcia-Betancur, J.C., Ramamurthi, K.S., Vlamakis, H., and Lopez, D. (2012). The biofilm formation defect of a *Bacillus subtilis* flotillin-defective mutant involves the protease FtsH. Mol Microbiol 86, 457-471. http://doi:10.1111/j.1365-2958.2012.08205.x

Yu, D., Zhao, L.P., Xue, T., and Sun, B.L. (2012). *Staphylococcus aureus* autoinducer-2 quorum sensing decreases biofilm formation in an *icaR*-dependent manner. Bmc Microbiol 12, 288. http://dx.doi.org/10.1186/1471-2180-12-288

Zafra, O., Lamprecht-Grandio, M., de Figueras, C.G., and Gonzalez-Pastor, J.E. (2012). Extracellular DNA Release by Undomesticated *Bacillus subtilis* Is Regulated by Early Competence. Plos One 7, e48716. https://doi.org/10.1371/journal.pone.0048716

Zuber, P. (2004). Spx-RNA polymerase interaction and global transcriptional control during oxidative stress. J Bacteriol 186, 1911-1918. http://dx.doi.org/10.1128/JB.186.7.1911-1918.2004

Mechanism of Biofilm Formation in Gram-negative Bacteria

2

Ulrich Vasconcelos*[1], Palashpriya Das*[2], Diogo Simas Bernardes Dias[3], Tarcísio Tárcio Correa Bonifácio[1], Ray Ravilly Alves Arruda[1], Bianca Teixeira Morais de Oliveira[1], Thiago Gonçalves Cavalcanti[1]

[1] Laboratório de Microbiologia Ambiental, Universidade Federal da Paraíba, João Pessoa-PB, Brasil

[2]National Institute of Pharmaceutical Education and Research (NIPER) Hajipur, India

[3]Centro de Tecnologia Mineral, Rio de Janeiro-RJ, Brasil

*Corresponding author: u.vasconcelos@cbiotec.ufpb.br and write2palashpryia@gmail.com

DOI: https://doi.org/10.21775/9781912530328.02

Abstract

Biofilms consist of one or more microbial species of microbes living in close association. Mixed species biofilms are more predominant in the environment. Though biofilm imparts various advantages to its producer, huge economic losses are encountered because of biofilm formation and much work has been described with special emphasis on *Pseudomonas aeruginosa*. However, studies have described beneficial effects of biofilms on human microbiota and that biofilm formation helps in maintenance of human health. Biofilm formation depends on physicochemical factors including temperature, oxygen, light, antibiotics and the interplay of the genes and their resultant products under various conditions. Since the first models on understandings of biofilms were evaluated in Gram-negative bacteria, the present review summarizes parameters that might be affecting this preferred form of microbial sustenance in different Gram-negative bacteria.

Introduction

Biofilms are composed of sessile microbial communities consisting of one or more species of microbes living in close association, surrounded by an adhesive polysaccharide matrix called exopolysaccharide (EPS). They remain adhered to a surface or an interface between two phases and are the preferred forms of microbial life on Earth (Santos et al. 2018). The first models on understandings of biofilms were evaluated in Gram-negative bacteria. Hence, in the present review, detailed mechanisms have been assessed which result in a microbial lifestyle within the community called biofilm.

Significance of Biofilms

The ability of bacteria to form communities attached to a substrate is of great importance in a myriad of disciplines, including biotechnology, engineering, industry, environmental sciences, materials science and medical sciences (Booth et al. 2011). In nature, biofilms provide many benefits like ensuring a prompt response and adaptations to environmental changes, allowing the establishment of a highly specialized and organized community, conferring stability to the structure and avoiding the exhaustion of nutrients and oxygen, as well as the accumulation of metabolites. Much is described about the problems related to infections, especially Gram-negative bacteria, with emphasis on *Pseudomonas aeruginosa* in approximately 80% of them (Brackman and Coenye, 2015).

Biofilms are associated with the transmission of pathogens through water pipes in hospitals and residences (McBain et al. 2003), orthoses (Kackar et al. 2017) and reduction of neutrophil phagocytosis resulting in reducing the efficacy of antimicrobial agents (Murphy and Caraher, 2015). In addition, biodegradation processes of different materials, such as metals (Li et al. 2004) and concrete (Das et al. 2016) as well as food spoilage (Speranza and Corbo, 2017) are also attributed to biofilms. However, biofilm formation can also be advocated for as they are important for human health with persistent cells establishing symbiosis in specific human niches (Renner and Weibel, 2011).

Many microorganisms are considered to be social beings. This characteristic feature can be attributed to the fact that they establish themselves as sessile multicellular communities, whose formation and maintenance are linked to different complex cellular and molecular mechanisms (Joo and Otto, 2012). It is believed that bacteria constitute a less preferred form of life in the planktonic state and approximately 99% of the bacteria live as sessile communities (Zhang, 2017; Garrett et al. 2008). However, some studies on microbial ecology suggest that a biofilm is a planktonic-like nursery that temporarily produces and protects cells, which are then released into the fluids and delivered to the environment through a liquid flow (Emerenini et al. 2015).

The biofilm can also be compared to a stationary phase culture. However, the population balance occurs in a natural state, different from those observed when bacterial cells are cultured in a synthetic media. The "natural stationary phase" is controlled by a set of intrinsic and extrinsic factors like oxygen consumption and nutrient enhancement, even when a considerable number of persistent cells occurs (Fricks-Lima et al. 2011). Hypothetically, there is no organic or inorganic surface capable of preventing the formation and development of a biofilm (Costerton, 1995). Thus, the ability to form such structures on any surface is solely a bacterial attribute, in which the mechanisms of recognition of the conditioning substrate and the processes involved in adhesion may vary (López et al. 2010).

Studies on Biofilms: Proposition of various models

The pioneering scientific study on biofilms was by Henrich in 1933 (Karunakaran et al. 2011). It was later deepened by the studies of ZoBell in the 1930s and 1940s, where the basis of Gram-negative marine bacterial affinity to certain surfaces was described (Costerton, 1995). Three decades later, a better understanding of the process of biofilm formation was delineated through the use of early mathematical models (Emerenini et al. 2015). Initially the models investigated a single species in a single conditioning medium, essentially considering the mass transport phenomena. Studies involving multispecies models were carried out in the 1980s. In the 1990s, multi-dimensional studies were performed using multispecies, multi-substrates, morphology, mechanisms and ecological interactions (Zhang, 2017). The 1990s also brought to light the non-random cellular organization in biofilms, proposed from the difficulty in understanding the concept of microbial life in an organized community when one believed in the planktonic state of life as the preferred form for microbes (Costerton and Wilson, 2004). This concept has brought advances in biofilm studies, bringing to the 21st century studies addressing spatial behaviors of the biofilm, using three-dimensional models with cyanobacteria (Clareli et al. 2016).

The current model of biofilms is based on *Pseudomonas aeruginosa*, an organism that became a standard through the large number of studies on the subject inspiring the creation of the term 'socio-microbiology', proposed in 2005 as a new area of bacteriology dedicated to the study of biofilms (Passos da Silva et al. 2017). The present biofilm model exhibits a complex three-dimensional dynamic structure composed of numerous microenvironments, where different osmotic gradients, nutrient concentrations and cell densities are established (Fux et al. 2005). This heterogeneity produces a number of advantages in allowing the establishment of different phenotypic varieties in an individual biofilm (Levipan and Avedaño-Herrera, 2017).

In addition, metabolic interactions are facilitated by proximity between cells thereby aiding in increased horizontal gene transfer and protection against environmental actions as well as predation (Kirisits et al. 2005). One report demonstrated this variety in a sessile structure of *P. aeruginosa* 57RP where individuals formed colonies of different sizes, exhibited distinct forms of mobility and presented rapid aggregation compared to other strains (Déziel et al. 2001). This heterogeneity produces a number of advantages in allowing the establishment of different phenotypic varieties in an individual biofilm (Levipan and Avedaño-Herrera, 2017). In addition, metabolic interactions are facilitated by proximity between cells thereby aiding in increased horizontal gene transfer and protection against environmental actions as well as predation (Kirisits et al. 2005).

Stages of biofilm formation

The process of biofilm formation involves different mechanisms of signaling and cell-dependence that can define the entire structure as a single organism (Brackman and Coenye, 2015). The life cycle of a biofilm can be divided into various stages and the migration of planktonic cells is considered as the first one. The adhesion stage, which can be reversible or permanent, is due to the surface recognition by the microorganism. The maturation involves a complex system of cellular communication, with EPS expression and flagellin suppression, culminating with the detachment, regulated by a number of factors, the result of which will be release of new planktonic cells in the medium (Emerenini et al. 2015; Sauer et al. 2002).

Migration and adhesion

The transport of the cells from a fluid flow to the conditioning surface has been considered by some to be the first stage in the formation of biofilm. This occurs as a function of a flow whose velocity influences how the cells will adhere and remain adhered to the substrate (Busscher et al. 1995). Until the beginning of adhesion, transport of the cells involves diffusion, convection and activation of flagella (Harmsen et al. 2010).

The adhesion process ensures that the microbes are not easily desorbed from the substrate (Marshal et al. 1971). The adhesion stage in Gram-negative bacteria is well reported in literature and the different mechanisms like adhesins, eDNA, flagella and pili with which this process is achieved have been described using *Pseudomonas aeruginosa* or *Escherichia coli* or *Salmonella* sp. as models (Skariyachan et al. 2018; Wang et al. 2016). This stage is subdivided into two phases, reversible and irreversible. In the first one, the sorption is not complete, and being regulated mainly by the limitation of nutrients or by the presence of inhibitors (Costerton, 1995). In case of irreversible adhesion, the process occurs rapidly, involving complex mechanisms of cell signaling resulting in the formation of microcolonies.

Maturation and maintenance of biofilm

A biofilm is composed of 85% matrix and 15% cells (Costerton and Wilson, 2004) and more than 90% of the dry mass is EPS (Toyofuko et al. 2015). The production and secretion of different polymers, mostly exopolysaccharides, gives *P. aeruginosa* the ability to establish itself in the preferred form of biofilms. In view of this, the bacterium is used as a model to understand the expansion of a biofilm.

In the maturation stage two types of populations occur. The first population comprises of vegetative cells producing autoinducers and the second is composed of differentiated flagellated cells responsible for the movement of the whole group (Schembri et al. 2002). The biofilm is formed progressively, connected by structures with a size smaller than 10 μm. A mature biofilm of *P. aeruginosa* reaches an elevation of up to 100 μm in 6 days, composed of non-motile cells, mostly segregated in clusters. In these conditions of protection by the EPS, the number of persistent cells begins to increase and at the 9th day, the first mobile cells involved in a less dense matrix are observed (López et al. 2010; Sauer et al. 2002). Cell density-dependent communication systems also participate in biofilm maturation. The maintenance stage regulates cell distribution and influences the architecture of the matrix. This promotes formation of holes in the interior in which few cells float.

Detachment

The biofilm detachment stage is considered by many authors to be the last stage of the biofilm formation and maintenance cycle. However, there has been an emerging notion that the detachment does not mark the end of the cycle because the transformation of non-motile cells into planktonic cannot be considered a final stage (Toyofuko et al. 2015). The detachment results in active dispersion regulated by genes controlling motility, chemotaxis and degradation of EPS. This ensures great phenotypic activity, significant physiological changes and production of lyases (Petrova and Sauer, 2016). As in other stages, cell dispersion and colony size are also regulated by complex mechanisms of quorum sensing signaling, but the rate of biofilm lysis occurs simply (Lamppa and Griswold, 2013).

Mathematical models with Gram-negative bacteria have determined that the detachment is induced by signaling initiated by factors external or internal to the biofilm (Emerenini et al. 2015). The release of cells is not merely a passive event of erosion or sloughing, but an active dispersive seeding process marked by events mediated by a complex metabolic structure regulated by EPS disrupting enzymes, production of surfactants, variation of nutrient concentration, ROS production and control by cell signaling (Emerenini et al. 2015). In order to adhere and colonize new conditioning surfaces in other habitats, bacteria return to the planktonic state as single cells or in small groups, but the death rate may be high because of lack of nutrients, competition with others microbes and the presence of possible predators (Brackman and Coenye, 2015). The escape strategies of cells of a biofilm can be characterized as forms of preservation and dissemination, whose objective is to avoid stresses associated with the development of the biofilm or to respond to selective aggressions offered by the environment.

Physicochemical factors affecting biofilm formation in Gram-negative bacteria

Cell-surface and cell-cell interactions, that are needed for the whole chain of events in biofilm formation to occur, is affected by different physicochemical factors (Karunakaran et al. 2011). The greatest limitation in community life involves the variation of nutrients in the environment and the presence of toxic metabolites. However, these conditions may interfere and regulate different responses that culminate in the cell's own persistence (Kirisits et al. 2005).

The adhesion process is dependent on the available carbon source ensuring the installation of the pioneer community (Berlanga and Guerrero, 2016). The time taken for this process can vary from within two hours in *Escherichia coli* (Balagué et al. 2011) to up to four hours in *Salmonella* sp. (Hover et al. 2016) reaching the state of mature biofilms at around 24 hours. The model of the *P. aeruginosa* biofilm architecture based on nutrient limitation was proposed few decades ago and can be described in two ways (Lebeaux et al. 2013). The model initially proposed, called the finger, is characterized by an architecture of pillars, while the new model presents the organization in a mushroom format. The available carbon source promotes changes in the shape assumed by the biofilm. Glucose allows the mushroom shape to be installed by promoting greater phenotypic heterogeneity of the cells. On the other hand, citrate leads to a uniform, flat and densely packed biofilm (Klausen et al. 2003). The structure also exhibits an iron dependence and under anoxic conditions, the architecture becomes more elongated (Harmsen et al. 2010). Iron also exerts effects in bacteria like *Flavobacterium psychrophilum* where iron-limiting conditions makes the bacterium deficient in biofilm formation (De la Fuente et al. 2013). The model of mushroom-like shape, joined by channels, ensures stability through the exchange of nutrients, metabolites and oxygen, involving a sophisticated signaling system (Monds and O'Tolle, 2009). In presence of oxygen, the mushroom presents channels and spaces between the macrocolonies with the objective of realizing exchanges (Toyofuko et al. 2015). The number of the pioneer cells forming the initial microcolonies also participate in the composition of the mushroom biofilm architecture (Ghanbari et al. 2016). Non-motile forms settle in the microcolonies, in the lower part of the mushroom and do not grow in number. The stalk thus exhibits a suppression of the contraction of the flagellum or the strong aggregation of cells creates an intracellular matrix which impedes motility. A motile subgroup cells colonizes the stalk and randomly migrates to the top, which extends through cell growth, as well as quorum sensing signals allowing loosening of the alginate, in a process similar to that in detachment. Hydrodynamic properties like fluid flow ratio as well as mobility mediated by the type IV pili expression, contribute to translocation and also widening of the upper part of the stalk. It is emphasized that this event depends on the density of the community in the stalk as well as availability of nutrients that favors chemotaxis (Klausen et al. 2003).

The ability of the cell to produce adhesion factors is crucial for the formation of a biofilm (Berne et al. 2015). Surface recognition is mediated by pili through the interactions between molecules present on the substrate, adhesins and pilins. Using this mechanism *E. coli* recognizes glycoproteins on the surface by means of type I pili (Schembri et al. 2002). Type IV pili plays an important role in transformation of the planktonic form of the cells to the sessile form. The pilin subunits are sintered as pre-pilins type A and B, processed prior to incorporating into the pilin structure (Giltner et al. 2012). The flp subunit of pre-pilin B is associated with adherence to the conditioning surface. Studies with *Vibrio vulnificus* have identified that the deletion of the Tad pili (*flp*) results in defective adhesion as well as in reduction of microbial resistance when the biofilm is established (Pu and Rowe-Magus, 2018). Additionally, fimbrial adhesins and pili are also involved in irreversible adhesion as was observed using *E. coli* as a model. These structures facilitate intracellular interactions during the aggregation and formation of microcolonies, with emphasis on pili type IV in *P. aeruginosa*, chaperone-usher pili (CUP) in enterobacteria, pili type I in *E. coli*, alternative CUP well characterized in Gram-negative bacteria and pilus class 5 and coli surface antigens (CS) in enterotoxigenic *E. coli* (ETEC), involved in the adhesion of the bacteria to the intestinal lumen (Karunakaran et al. 2011).

The topography of the conditioning surface will guide how adhesion can occur (Renner and Weibel, 2011). Bacteria are negatively charged and adhere to positively charged surfaces or are repelled by contact with surfaces of the same charge but fimbriae can minimize this destabilization. In addition, hydrophilic surfaces require more energy for adhesion and preferably the cells adhere to hydrophobic ones (Yuan et al. 2017). Interestingly, strains with more R plasmids adhere more frequently than those that have less or none. The mechanism of how this occurs has not yet been fully understood, but it is believed to be related to the fact of encoding different surface components (Oufrid et al. 2014). The adhesion of particles smaller than 1 μm can be explained based on the behavior of a colloid as predicted by Derjaguin-Landau and Verwey-Overbeek theory, DLVD (Oliveira, 1997). This theory involves the estimation of the magnitude and variation of the inter-particular distance of a London-Van der Waals attraction energy between two surfaces and the repulsive energy resulting from the overlapping of ionic atmospheres between them (Marshal et al. 1971). Thus, cell adhesion on the conditioning surface is determined by the equilibrium between van der Waals forces due to double-layer electrical interactions. The ratio between the surface area and the volume of the microbial cell in relation to the substrate also contributes to the sorption as a consequence of long-range forces similar to electrostatic interactions and Van der Waals forces (Karunakaran et al.

2011). The surface composition may prevent the diffusion of autoinducers and thus repress the mechanisms used in the stages of planktonic cell transformation and substrate recognition (Stewart, 2003).

Some molecules can inhibit the process of adhesion and even block it completely. Sub-inhibitory concentrations of pyocyanin, may reduce the formation of biofilms of *E. coli* and *Enterobacter aerogenes* without altering the concentration of planktonic cells of enterobacteria. This may be a strategy of the producer to migrate to an appropriate conditioning surface avoiding competition (Viana et al. 2017). The literature also reports other cases of molecules with antimicrobial properties inhibiting the adherence of *E. coli* when applied in sub-inhibitory concentrations, as in the case of cefotaxime (Balagué et al. 2011). Hydrophilic materials may also inhibit the adhesion of some cells.

The alteration in the membrane surface of the bacterium is involved with the adhesion and subsequent formation of the biofilm. While inverse correlation between cell surface hydrophobicity and biofilm formation has been depicted in *Bradyrhizobium japonicum* (Park and So, 2002), positive correlation was detected in *Stenotrophomonas* sp. (Pompilio et al. 2008). Again, bacterial strains of *Proteus mirabilis* with more negative value of zeta potential, formed more amount of biofilm than the strains with less negative values of zeta potential (Czerwonka et al. 2016). The hydrophilic membranes of the *E. coli* 3519EC1 and DH10B strains do not allow the bacteria to adhere to carbon steel surfaces. When the first one was treated with a resistance plasmid, the membrane became more hydrophobic resulting in biofilm formation. Thus, it can be inferred that as a strategy of growth, intercellular exchanges can occur between organisms in close proximity, ensuring community life, protection against environmental stresses, as well as from exerted selective pressures (Oufrid et al. 2014).

The collapse of the biofilm structure is linked to two mechanisms: reduction of c-di-GMP and quorum sensing signaling. Models using *P. aeruginosa* showed that the increase in c-di-GMP concentration promoted the growth of sessile cells whereas induction of transformations to the planktonic phenotype occurred conversely. c-di-GMP levels are modulated by three molecules: diguanylate cyclases (DGC), proteins with GGDEF domain and phosphodiesterases (PDEs) (Petrova and Sauer, 2016). The second mechanism identified in *P. aeruginosa* is subtly more complex and involves the accumulation of autoinducers that regulate the activity of periplasmic tyrosine phosphatase (TbbA) through LasI/RhRR systems. This culminates in membrane protein dephosphorylation, thus reducing c-di-GMP and PEL levels via inhibition of PelD. These same systems still promote the synthesis of rhamnolipids which, due to their tenso-active properties, coupled with the reduction of the rheological properties of the polymeric matrix, favor the detachment (Solano et al. 2014). In this context, the reduction of biofilm adhesiveness and matrix disruption favoring the emigration of flagellate cells is a necessary event for the survival of the biofilm, emphasizing once again that the availability of nutrients plays an important role in the events that culminate with the release of the cells (Harmsen et al. 2010). In the context of detachment, three pathways of cell output in a biofilm are identified (Stoodley et al. 2002). The first of these is dispersion governed by different internal and external factors such as oxygen depletion, presence of peroxides, reduction and extinction of fluid flow, sudden changes in assimilable carbon concentration, pH and the presence of heavy metals. The second route is the release which is the result of degradation of the matrix involving enzymes such as alginate lyases and surfactants such as dispersin B. The third and last mechanism is via desorption induced by the same factors that lead to dispersion as well as by the presence of nitric oxide and proteinase K (Petrova and Sauer, 2016).

The establishment of a biofilm has a significant correlation with temperature. The increase in temperature is associated with a greater availability of nutrients, as well as an increase in the natural adhesion of the cells, which allows the installation of biofilms under temperatures between 80–90 ºC, an event known as baking effect (Garrett et al. 2008). On the other hand, the reduction of the temperature contributes to the decrease of the adhesiveness of the polymer on a conditioning surface, possibly due to the change in the relation between the contact area between the polymer, bacterial area and the adhering surface (Zeraik and Nitschke, 2012). pH plays an important role in biofilm stability, since

EPS production occurs preferentially at neutral pH. In addition, autoinducers related to cell signaling processes are less sensitive at weakly acidic and neutra pH, while in the alkaline range, the half-life of the autoinducers exhibits a drop for minutes (Hense and Schuster, 2015).

Campylobacter jejuni is a microaerophilic microorganism but this strain is able to form a structured biofilm when cultured in aerobic condition. This behavior has been speculated to be resulting from modification in the regulation of severa genes involved in oxidative stress response, oxygen metabolism, adhesion, and biofilm formation (Bronnec et al. 2016). Light is a major factor for the formation of biofilms in cyanobacteria. Modeling studies have revealed that light has a significant effect on adhesion events, mainly, but temperature and nutrients need to be present to ensure the establishment of biofilms in cyanobacteria (Clarelli et al. 2016).

Heterogenicity of cells in the biofilm, not present in planktonic individuals, has been reported to be promoted by environmental factors, especially changes in the pH gradient, as well as oxygen and nutrient concentrations (Booth et al. 2011). The elasticity of the matrix is guaranteed by the production of acetylated uronic acids that act as compatible solutes maintaining hydration of the structure, thereby stabilizing it. Other factors involved in matrix elasticity and plasticity are the nature of the polymer and the weak hydrogen bond (Garrett et al. 2008). Once stable, the biofilm architecture itself ensures an increase in resistance to a myriad of environmental stressors, which is absent in the planktonic population, the mostly studied stressor being resistance to antimicrobial agents (Booth et al. 2011). *P. aeruginosa* sessile biofilms, for example, exhibit 1000 times more antimicrobial resistance than planktonic bacterial cells (Fricks-Lima et al. 2011). Positively charged antibiotics have been reported to be slowed down in action on biofilms by the negative charge on the alginate matrix (Fux et al. 2005). This physical barrier generates a gradient of antibiotics across the membrane which leads to reduced concentrations at sub-inhibitory levels, ensuring development of resistance, formation of persistent subpopulations and increased horizontal transfer of the resistance factors (Toyofuko et al. 2015). The following table (Table 1) presents different examples of how environmental stresses can produce persistent subpopulations during a biofilm formation cycle.

Table 1. Responses and defense mechanisms of sessile cells of Gram-negative bacteria

Stress responses	Mechanisms	Reference
Antimicrobial Resistance	Reduction of penetration, growth rate & metabolism	Brackman and Coenye, 2015
Resistance to environmental stresses	Horizontal gene transfer	Kirisits et al. 2005
Resistance to ROS	Formation of persistent cells	López et al. 2010
Resistance to heavy metals	Persistent cell generation	Zhang, 2017
Reduction of macrophage activity	Alginate production	Gholami et al. 2017
Defense against phage and immune system	Alginate production	Skariyachan et al. 2018

Molecular genetics governing biofilm formation in Gram-negative bacteria

The formation of a biofilm is a complex process involving different genes (Kreth and Herzberg, 2015). Models with *P. aeruginosa* determined that the genes involved are responsible for the expression of multiple phenotypes at all stages of establishment and maturation of biofilm in the bacteria (Sauer et al. 2002). *P. aeruginosa* cells may exhibit different gene expressions but they do not show differences in gene composition (López et al. 2010). Overexpression of the *psl* and *pel* locus, for example, directly reflects the motility and EPS functions. This was well discussed in a comparative study of wild isolates of *P. aeruginosa* in which liquid self-aggregation, hyper-adhesion on solid substrates, increased hydrophobicity, reduced motility and changes in colony morphology were observed (Kirisits et al. 2005). The Cpx system is also responsible for the expression of cellular signals after adhesion, specifically in alteration within cellular envelope and in activation of genes encoding the plasma proteins, the degradation factors of which affect synthesis of alginate (Schembri et al., 2002). The formation of alginate by *P. aeruginosa* is encoded by *alg*D and *alg*U genes. While these genes are present in 80% and 70% of wild isolates respectively, they are present in 98% of bacterial isolates involved in outbreaks and/or cases of nosocomial infections involving the formation of biofilms (Gholami et al. 2017).

The entire process of biofilm formation is part of a cell density-dependent communication system in which chemical signals, called autoinducers, are produced, secreted and assimilated after which it coordinates different responses. The synthesis of autoinducers does not depend on the concentration of nutrients in the medium and are always being secreted at basal levels, with at least 500 cells (Emerenini et al. 2015; Brackman and Coenye, 2015). The correlation of the quorum sensing system and the biofilm formation of *P. aeruginosa* was described at the end of the decade of 1990 and from there, similar mechanism was identified in other Gram-negative organisms like *V. cholerae*, *Xanthomonas campestris* (Solano et al. 2014), *Burkholderia* sp. (Passos da Silva et al. 2017) and alpha and betaproteobacteria (Schembri et al. 2002). The quorum sensing systems in *Vibrio* sp. comprise autoinducers derived from S-adenosyl-methionine, SAM, as well as acyl homoserine lactones, AHL (Toyofuko et al. 2015).

The models of quorum sensing were investigated more frequently using *Pseudomonas aeruginosa* whose systems, formed by *las* and *rhl*, also participate in the AHL-type autoinducer synthesis, initially identified with the matrix production (López et al. 2010). There are three AHLs involved in this process: 3-oxo-C12-homoserine lactone from *las* system, C4-homoserine lactone from *rhl* system and 2-heptyl-3-hydroxy-4-quinolone coupled to the *pqs* system (Harmsen et al. 2010). The *las* quorum sensing system further regulates the process in which flagellated pseudomonad cell forms exhibit a typical rotational motion on the axis itself upon initiation of the adhesion process whereas in the non-motile strains of *Pseudomonas aeruginosa* PA257, adhesion occurs laterally thereby implying a cell strategy during the process reversal, as the laterally adhered cells are easier to desorb from the surface where they were initially adhered (Sauer et al. 2002).

The process of adhesion during biofilm formation in Gram-negative bacteria is also regulated by two other mechanisms. While the first mechanism is related to natural conjugation of plasmids inducing planktonic cell transformation into sessile cells (Oufrid et al. 2014), the other, described in *E. coli* K-12, involves expression of genes controlling transition from the planktonic to the sessile state. Additionally, at least four classes of proteins have been identified to be involved in this mechanism in *P. putida* which are responsible for expression of metabolites, in biosynthesis of lipids and membrane proteins, transport as well as events related to adaptation and protection of cells (Rühl et al. 2012).

The second messenger, bis-(3'-5'') cyclic dimeric guanosine monophosphate (c-di-GMP) is also important in transition from planktonic state to the sessile one. Higher amounts of c-di-GMP was found within the cells of biofilm as compared to planktonic cells of *Acidithiobacillus thiooxidans* (Diaz et al. 2018). c-di-GMP positively regulates biofilm maturation and negatively affects planktonic life in *Aeromonas* sp. Autoinducer-mediated quorum sensing is involved in the control of biofilm by c-di-GMP (Talagrand-Reboul et al. 2017). It plays a role in the production and secretion of the EPS matrix (Toyofuko et al. 2015). EPS produced by *P. aeruginosa* can be made up of different structures, the most common being alginate, whose expression is regulated by 12 genes (PA3540-PA3551). Alginate plays an important role as a ROS scavenger, participating in the defense of bacteria against environmental stresses (Skariyachan et al.

2018). Two other polymers are known, Pel (glucose-rich polymer), derived from the word pellicle and Psl (mannose-rich polymer composed of alternating repeated units of L-rhamnose, D-glucose and D-mannose). However, the expression of Pel or Psl varies according to the strain (López et al. 2010). Pel is a cationic EPS identified in *P. aeruginosa* PA14 and the *pel* gene operon products (PA3058-PA3064) control its synthesis. With less alginate, the bacterium uses the *pel* or *psl* system, with 15 genes, among which 11 are needed for its synthesis. In the absence of Psl, the matrix viscosity is increased (Skariyachan et al. 2018).

The cell-dependent communication system, quorum sensing, can also participate in alteration of responses and in the enzymatic metabolism of auto-inducers or quorum quenching based on mutational analysis of autoinducer-producing genes (Brackman and Coenye, 2015). Not only are they involved in alginate expression, natural quorum sensing systems also modulate cellular responses and control gene expression when the cells are still in the planktonic state. Bacteria live in fixed volumes where autoinducers are produced and accumulate. Flow velocity plays an important role in the repression of some responses by means of convective transport, as observed in *V. cholerae*. In environments with high flow, AHL dispersion increases and adhesion can be suppressed (Passos da Silva et al. 2017). In addition, under different flow regime, the responses obtained are distinct among distinct species. A study simulating synthetic quorum sensing of strains like *E. coli* and *P. aeruginosa* identified that the modulator systems were activated in different ways when the critical population was reached, due to the presence of flow regime. While the responses in wild type *P. aeruginosa* strains were repressed with a flow rate of 1000 μm/s, *E. coli* were activated up to a flow rate of 250 μm/s (Emge et al. 2016).

Other inducers not dependent on cell density may degrade the expression of genes in Gram-negative cells. In *P. aeruginosa* and *E. coli*, sub-inhibitory concentrations of antibiotics, rhamnolipids, eDNA, phenazine pigments and siderophores may participate in the cellular activation mechanisms. In *P. aeruginosa*, pyocyanin, a pigment exhibited by 90-95% of the strains has an important role of extracellular electron transfer to generate energy for growth, resulting in differentiation of the colonies by the induction of the *sox*R gene (Passos da Silva et al. 2017; Harmsen et al. 2010; Vasconcelos et al. 2010). Pyocyanin also participates in the release of eDNA, inducing lysis mediated by the production of hydrogen peroxide. The pigment then interacts with the nitrogenous bases of the DNA and creates a deformation in the structure of the double helix. The pyocyanin-eDNA complex interferes with cell surface hydrophobicity and cell-cell physicochemical interactions, creating conditions for the establishment of a robust biofilm. This effect can be reversed by glutathione which binds directly to pyocyanin modifying its structure and inhibiting its activity (Das et al. 2016).

Conclusion

This chapter discusses the mechanisms of biofilm formation in Gram-negative bacteria with focus on *Pseudomonas aeruginosa*. Many Gram-negative organisms can form biofilms in clinically relevant cases and hence an understanding of biofilm formation is required for understanding the microbial processes that are similar yet unique to a given microbial species. Unraveling a detailed understanding of the mechanisms of biofilm formation in Gram-negative bacteria might also develop newer strategies to combat biofilms in those conditions.

References

Balagué, C., Pérez, J., Rinaudo, M., and Fernández, L. (2011). Inhibition of the adhesion to catheters of uropathogenic *Escherichia coli* by sub-inhibitory concentrations of cefotaxime. Eur J Obstet Gynecol Reprod Biol. *155*, 150-156 DOI: 10.1016/j.ejogrb.2010.12.034.

Berlanga, M., and Guerrero, R. (2016). Living together in biofilms: the microbial cell factory and its biotechnological implications. Microb Cell Fact. *15*, 165 DOI: 10.1186/s12934-016-0569-5.

Berne, C., Ducret. A., Hardy, G. G., and Brun, Y. V. (2015). Adhesins involved in attachment to abiotic surfaces by Gram-negative bacteria. Microbiol Sprect. *3*, 1-27 DOI: 10.1128/microbiolspec.MB-0018-2015.

Booth, S. C., Workentine, M. L., Wen, J., Shaykhutdinov, R., Vogel, H. J., Ceri, H., Turner, R. J., and Weljie, A. M. (2011). Differences in metabolism between the biofilm and planktonic response to metal stress. J Proteme. *10*, 3190-3199 DOI: 10.1021/pr2002353.

Brackman, G., and Coenye, T. (2015). Quorum sensing inhibitors as anti-biofilm agents. Curr Pharm Design. *21*, 5-11 DOI: 10.2174/1381612820666140905114627.

Bronnec, V., Turoňová, H., Bouju, A., Cruveiller, S., Rodrigues, R., Demnerova, K., Tresse, O., Haddad, N., and Zagorec, M. (2016). Adhesion, Biofilm Formation, and Genomic Features of *Campylobacter jejuni* Bf, an atypical strain able to grow under aerobic conditions. Front Microbiol. *7*, 1002 DOI: 10.3389/fmicb.2016.01002.

Busscher, H. J., Bos, R., and Van der Mei, H. C. (1995). Initial microbial adhesion is a determinant for the strength of biofilm adhesion. FEMS Microbiol Lett. *128*, 229-234.

Clarelli, F., Russo, C., Natalini, R., and Ribot, M. (2016). A fluid dynamics multidimensional model of biofilm growth: stability, influence of environment and sensitivity. Math Med Biol. *33*, 371-395 DOI: 10.1093/imammb/dqv024.

Costerton, J. W. (1995). Overview on biofilms. J Ind Microbiol. *15*, 137-140.

Costerton, W. J., and Wilson, M. (2004). Introducing biofilms. Biofilms. *1*, 1-4.

Czerwonka, G., Guzy1, A., Kałuza K., Grosicka M., Danczuk M., Lechowicz L., Gmiter D., Kowalczyk P., and Kaca W. (2016). The role of *Proteus mirabilis* cell wall features in biofilm formation. Arch Microbiol. *198*:877-884.

Das, T., Ibugo, A. I., Klare, W., and Manefield, M. (2016). Role of pyocyanin and extracellular DNA in facilitating *Pseudomonas aeruginosa* biofilm formation. In Microbial biofilms – importance and applications, eds. D. Dhanasekaran and N. Thajuddin N., (London: ItechOpen), pp. 23-42 DOI: 10.5772/63497.

De la Fuente, M., Vidal, J. M., Miranda, C. D., González, G., and Urrutia, H. (2013). Inhibition of *Flavobacterium psychrophilum* biofilm formation using a biofilm of the antagonist *Pseudomonas fluorescens* FF48. Springerplus. *2*, 176 DOI: 10.1186/2193-1801-2-176.

Déziel, E., Comeau, Y., and Villemur, R. (2001). Initiation of biofilm formation by *Pseudomonas aeruginosa* 57RP correlates with emergence of hyperpiliated and highly adherent phenotypic variants deficient in swimming, swarming, and twitching motilities. J Bacteriol. *183*, 1195-1204 DOI: 10.1128/JB.183.4.1195-1204.2001.

Díaz M., Castro M., Copaja S., and Guiliani, N. (2018). Biofilm formation by the acidophile bacterium *Acidithiobacillus thiooxidans* involves c-di-GMP pathway and *pel* exopolysaccharide. Genes. *9*, 113 DOI: 10.3390/genes9020113.

Emerenini, B. O., Hense, B. A., Kuttler, C., and Eberl, H. J. (2015). A mathematical model of quorum sensing induced biofilm detachment. PLOS One. *10*, e0132385 DOI: 10.1371/journal.pone.0132385.

Emge, P., Moeller, J., Jang, H., Rusconi, R., Yawata, Y., Stocker, R., and Vogel, V. (2016). Resilience of bacterial quorum sensing against fluid flow. Sci Rep. *6*, 33115 DOI: 10.1038/srep33115.

Fricks-Lima, J., Hendrickson, C. M., Allgaier, M., Zhuo, H., Wiener-Kronish, J. P., Lynch, S. V., and Yang, K. (2011). Differences in biofilm formation and antimicrobial resistance of *Pseudomonas aeruginosa* isolated from airways of mechanically ventilated patients and cystic fibrosis patients. Int J Antimicrob Agents. *37*, 309-315 DOI: 10.1016/j.ijantimicag.2010.12.017.

Fux, C. A., Costerton, W. J., Stewart, P. S, and Stooley, P. (2005). Survival strategies of infectious biofilms. Trends Microbiol. *13*, 34-40 DOI: 10.1016/j.tim.2004.11.010.

Garrett, T.R., Bhakoo, M., and Zhang, Z. (2008). Bacterial adhesion and biofilm surfaces. Progress Nat Sci. *18*, 1049-1056 DOI: 10.1016/j.pnsc.2008.04.001.

Ghanbari, A., Dehghany, J., Schwebs, T., Müsken, M., Häussler, S., and Meyer-Hermann, M. (2016). Inoculation density and nutrient level determine the formation of mushroom-shaped structures in *Pseudomonas aeruginosa* biofilms. Sci Rep. *6*, 32907 DOI: 10.1038/srep32097.

Gholami, S., Tabatabaei, M., and Sohrabi, N. (2017). Comparison of biofilm formation and antibiotic resistance pattern of *Pseudomonas aeruginosa* in human and environmental isolates. Microb Pathog. *109*, 94-98 DOI: 10.1016/j.micpath.2017.05.004.

Giltner, C. L., Nguyen, Y., and Burrows, L. L. (2012). Type IV pilin proteins: versatile molecular modules. Microbiol Mol Biol Rev. *76*, 740-772 DOI: 10.1128/MMBR.00035-12.

Harmsen, M., Yang, L., Pamp, S. J., and Tolker-Nielsen, T. (2010). An update on *Pseudomonas aeruginosa* biofilm formation, tolerance, and dispersal. FEMS Immunol Med Microbiol. *59*, 253-268 DOI: 10.1111/j.1574-695X.2010.00690.x.

Hense, B. A., and Schuster, M. (2015). Core principles of bacterial autoinducer systems. Microbiol Mol Biol Rev. *79*, 153-169 DOI: 10.1128/MMBR.00024-14.

Hover, T., Maya, T., Ron, S., Sandovsky, H., Shadkchan, Y., Kijner, N., Mitiagin, Y., Fichtman, B., Harel, A., Shanks, R. M. Q., Bruna, R. E., García-Véscovi, E., and Osherov, N. (2016). Mechanisms of bacterial (*Serratia marcescens*) attachment to, migration along, and killing of fungal hyphae. Appl Environ Microbiol. *82*, 2585-2594 DOI: 10.1128/AEM.04070-15.

Joo, H-S., and Otto, M. (2012). Molecular basis of *in-vivo* biofilm formation by bacterial pathogens. Chem Biol. *19*, 1503-1513 DOI: 10.1016/j.chembiol.

Kackar, S.; Suman, E., and Kotian, M. S. (2017). Bacterial and fungal biofilm formation on contact lenses and their susceptibility to lenses care solutions. Indian J Med Microbiol. *35*, 80-84 DOI: 10.4103/ijmm.IJMM_16_273.

Karunakaran, E., Mukherjee, J., Ramalingam, B., and Bigss, C. A. (2011). "Biofilmology": a multidisciplinary review of the study of microbial biofilms. Appl Microbiol Biotechnolol. *90*, 1869-1881 DOI: 10.1007/s00253-011-3293-4.

Kirisits, M. J., Prost, L., Starkey, M., and Parsek, M. R. (2005). Characterization of colony morphology variants isolated from *Pseudomonas aeruginosa* biofilms. Appl Environ Microbiol. *71*, 4809-4821 DOI: 10.1128/AEM. 71.8.4809-4821.2005.

Klausen, M., Aaes-Jørgensen, A., Molin, S., and Tolker-Nielsen, T. (2003). Involvement of bacterial migration in the development of complex multicellular structures in *Pseudomonas aeruginosa* biofilms. Mol Microbiol. *50*, 61-68.

Kreth, J., and Herzberg, M. C. (2015). Molecular principles of adhesion and biofilm formation. In The root canal biofilm – Series on biofilm 9, L. E. Chávez de Paz, C. M. Sedgley, A. Kisthen, ed. (New York: Springer), pp. 23-53 DOI: 10.1007/978-3-662-47415-0_2.

Lamppa, W., and Griswold, K. E. (2013). Alginate lyase exhibits catalysis-independent biofilm dispersion and antibiotic synergy. Antimicrob Agents Chemother. *57*, 137-145. DOI: 10.1128/AAC.01789-12.

Lebeaux, D., Chauan, A., Rendueles, O., and Beloin, C. (2013). From *in vitro* to *in vivo* models of bacterial biofilm-related infections. Pathogens. 2, 288-356 DOI:10.3390/pathogens2020288.

Levipan, H. A., and Avendaño-Herrera, R. (2017). Different phenotypes of mature biofilm in *Flavobacterium psychrophilum* share a potential for virulence that differs from planktonic state. Front Cell Infect Microbiol. *15*, 76 DOI: 10.3389/fcimb.2017.00076.

Li, B., and Logan, B. E. (2004). Bacterial adhesion to glass and metal-oxide surfaces. Colloids Surf B. *36*, 81-90 DOI: 10.1016/j.colsurfb.2004.05.006.

López, D., Vlamakis, H., and Kolter, R. (2010). Biofilms. Cold Spring Harb Perspect Biol. 2, a000398 DOI: 10.1101/cshperspect.a000398.

Marschall, K. C., Stout, R., and Mitchell, R. (1971). Mechanism of the initial events in the sorption of marine bacteria to surfaces. J Gen Microbiol. *68*, 337-348.

McBain, A. J., Bartolo, R. G., Catrenich, C. E., Charbonneau, D., Ledder, R. G., Rickard, A. H., Symmons, S. A., and Gilbert, P. (2003). Microbial characterization of biofilms in domestic drains and the establishment of stable biofilm microcosms. Appl Environ Microbiol. *69*, 177-185.

Monds, R. D., and O'Toole, G. A. (2009). The developmental model of microbial biofilms: ten years of a paradigm up for review. Trends Microbiol. *17*, 73-87 DOI: 10.1016/j.tim.2008.11.001.

Murphy, M. P. and Caraher, E. (2015). Residence in biofilms allows *Burkholderia cepacia* complex (Bcc) bacteria to evade the antimicrobial activities of neutrophil-like dHL60 cells. FEMS Pathog Dis, *73*, ftv069 DOI: 10.1093/femspd/ftv069.

Oliveira, R. (1997). Understanding adhesion: A means for preventing fouling. Exp Thermal Fluid Sci. *14*, 316-322 DOI: 10.1016/S0894-1777(96)00134-3.

Oufrid, S., Mliji, E. M., Latrache, H., Mabrouki, M., Hamadi, F., Talmi, M., and Timinouni, M. (2014). Plasmid mediated resistance to cephalosporin and adhesion properties in *E. coli.* I J Eng Res Appl. *4,* 55-62 DOI: 10.1111/1462-2920.13722.

Park, K. M. and So, J. S. (2000) Altered cell surface hydrophobicity of lipopolysaccharide-deficient mutant of *Bradyrhizobium japonicum.* J Microbiol Methods. *41*, 219–226 DOI: 10.1016/ S0167-7012(00)00155-X.

Passos da Silva, D., Schofield, M. C., Parsek, M. R., and Tseng, B. S. (2017). An update on the sociomicrobiology of quorum sensing in Gram-negative biofilm development. Pathogens. *6*, 51 DOI:10.3390/pathogens6040051.

Petrova, O. E., and Sauer, K. (2016). Escaping the biofilm in more than one way: desorption, detachment or dispersion. Curr Opin Microbiol. *30*, 67-78 DOI: 10.1016/j.mib.2016.01.004.

Pompilio, A., Piccolomini, R., Picciani, C., D'Antonio, D., Savini, V., and Di Bonaventura, G. (2008). Factors associated with adherence to and biofilm formation on polystyrene by *Stenotrophomonas maltophilia*: the role of cell surface hydrophobicity and motility. FEMS Microbiol Lett. *287*, 41–47 DOI: 10.1111/j.1574-6968.2008.01292.x.

Pu, M., and Rowe-Magnus, D. A. (2018). A Tad pilus promotes the establishment and resistance of *Vibrio vulnificus* biofilms to mechanical clearance. NPJ Biofilm Microbes. *4*, 10 DOI: 10.1038/s41522-018-0052-7.

Renner, L. D., and Weibel, D. B. (2011). Physicochemical regulation of biofilm formation. MRS Bull. *36*, 347–355 DOI: 10.1557/mrs.2011.65.

Rühl, J., Hein, E-M., Hayen, H., Schmid, A., and Blank, L. M. (2012). The glycerophospholipid inventory of *Pseudomonas putida* is conserved between strains and enables growth condition-related alterations. Microb Biotechnol. *5*, 45-58 DOI: 10.1111/j.1751-7915.2011.00286.x.

Santos, A. L. S., Galdino, A. C. M., Mello, T. P., Ramos, L. S., Branquinha, M. H., Bolognese, A. M., Columbano Neto, J., and Roudbary, M. (2018) What are the advantages of living in a community? A microbial biofilm perspective! Mem Inst Oswaldo Cruz. *113*, 1-7 DOI: 10.1590/0074-02760180212.

Sauer, K., Camper, A. K., Ehrlich, G. D., Costerton, J. W., and Davies, D. G. (2002). *Pseudomonas aeruginosa* displays multiple phenotypes during development as a biofilm. J Bacteriol. *184*, 1140-1154.

Schembri, M. A., Givskov, M., and Kelmm, P. (2002). An attractive surface: Gram-negative bacterial biofilms. Sci STKE. *14*, r06 DOI: 10.1126/stke.2002.132.re6.

Skariyachan, S. Sridhar, V. S., Packirisamy, S., Kumargowda, S. T., and Challapilli, S. B. (2018). Recent perspectives on the molecular basis of biofilm formation by *Pseudomonas aeruginosa* and approaches for treatment and biofilm dispersal. Folia Microbiol. *63*, 413-432 DOI: 10.1007/s12223-018-0585-4.

Solano, C., Echeverz, M., and Lasa, I. (2014). Biofilm dispersion and quorum sensing. Curr Opin Microbiol. *18*, 96-104 DOI: 10.1016/j.mib.2014.02.008.

Speranza, B., and Corbo, M. R. (2017). The Impact of Biofilms on Food Spoilage. In: The Microbiological Quality of Food, eds. A. Bevilacqua, M. R. Corbo and M. Sinigaglia. (Amsterdam: Elsevier), pp. 259-282 DOI: 10.1016/B978-0-08-100502-6.00014-5.

Stewart, P. S. (2003). Diffusion in Biofilms. J. Bacteriol. *185*, 1485-1491. DOI: 10.1128/JB.185.5.1485–1491.

Stoodley, P., Sauer, K., Davies, D. G., and Costerton, J. W. (2002). Biofilms as complex differentiated communities. Annu Rev Microbiol. *56*, 187-209 DOI: 10.1146/annurev.micro.56.012302.160705.

Talagrand-Reboul, E., Jumas-Bilak, E., and Lamy, B. (2017). The social life of *Aeromonas* through biofilm and quorum sensing systems. Front Microbiol. *8*, 37 DOI: 10.3389/fmicb.2017.00037.

Toyofuko, M., Inaba, T., Kiyokawa, T., Obana, N., Yawata, Y., and Nomura, N. (2015). Environmental factors that shape biofilm formation. Biosci Biotechnol Biochem. *80*, 7-12 DOI: 10.1080/09168451.2015.1058701.

Vasconcelos, U., Lima, M. A. G. A., and Calazans, G. M. T. (2010). *Pseudomonas aeruginosa* associated with negative interactions on coliform bacteria growth. Can J Pure Appl Sci. *4*, 1133-1139.

Viana, A. A. G., Martins, R. X., Ferreira, G. F., Zenaide-Neto, H., Amaral, I. P. G., and Vasconcelos, U. (2017). *Pseudomonas aeruginosa* and pyocyanin negatively act on the establishment of Enterobacteriaceae biofilm on a ceramic surface. Int J Eng Res Appl. *7*, 23-30 DOI: 10.9790/9622-0708022330.

Wang, H., Dong, Y., Wang, G., Xu, X., and Zhou, G. (2016). Effect of growth media on gene expression levels in *S. typhimurium* biofilm formed on stainless steel surface. Food Control. *59*, 546-552 DOI:10.1016/j.foodcont.2015.06.026.

Yuan, Y., Hays, M. P., Hardwidge, P. R., and Kim, J. (2017). Surface characteristics influencing bacterial adhesion to polymeric substrates. RSC Adv. *7*, 14254-14261 DOI: 10.1039/c7ra01571b.

Zeraik, A. E., and Nitschke, M. (2012). Influence of growth media and temperature on bacterial adhesion to polystyrene surfaces. Braz Arch Biol Technol. *55*, 569-576 DOI: 10.1590/S1516-89132012000400012.

Zhang, T. Modeling biofilms: From genes to communities. (2017). Processes. *5*, 5 DOI: 10.3390/pr5010005.

Existing and Novel Techniques to Study Biofilms

3

Paramita Basu*
Department of Pharmaceutical & Biomedical Sciences, Touro college of Pharmacy, New York, NY, USA

*Corresponding author: paramita.basu@touro.edu

DOI: https://doi.org/10.21775/9781912530328.03

Abstract

Bacterial biofilms are aggregates of microorganisms in which cells adhere to each other and also to a solid surface or an animal host cavity (oral cavity in humans and rats and nasal cavities in non-human hosts). Biofilms also commonly occur on the surface of in-dwelling devices in humans, on dentures and medical devices and sutures etc. placed inside the body. The complex composition of bacterial biofilms with EPS (extracellular polymeric substance) includes proteins, polysaccharides, extracellular DNA, peptidoglycan, lipids and phospholipids. These substances play a role in the initial adhesion of bacteria to a surface and maintenance of the biofilm structure. This chapter describes various traditional and novel techniques that are available for studying structures of bacterial biofilms, their analysis and quantification. The methods vary in their approach and include colorimetric and fluorescence-based methods to evaluate and quantify biofilm matrix and viability, direct imaging using different types of microscopy, as well as molecular approaches including omics studies and are described in terms of methodology and application.

Introduction

Biofilms, which have multiple definitions, but generally refer to communities of microbes (prokaryotic and eukaryotic) that adhere to a surface and are enclosed in an extracellular polysaccharide matrix (Heukelekian et al., 1940; Costerton et al.,1987; Costerton et al.,2004). Van Leeuwenhoek, using his simple microscopes, first described microorganisms on tooth surfaces and can be credited with the discovery of microbial biofilms. Heukelekian and Heller observed the "bottle effect" for marine microorganisms, i.e., bacterial growth and activity were substantially enhanced by the incorporation of a surface to which these organisms could attach. This microbiological phenomenon, that microorganisms attach to and grow universally on exposed surfaces led to studies that revealed surface-associated microorganisms (biofilms) exhibited a distinct phenotype with respect to gene transcription and growth rate. These biofilm microorganisms have been shown to elicit specific mechanisms for initial attachment to a surface, development of a community structure or ecosystem and detachment.

Light microscopy, transmission and scanning electron microscopy (Costerton, 2004; Marrie, 1982; Long, 1972; Kinner, 1985) played a pivotal role in the realization of the biofilm concept. Through the application of transmission electron microscopy, morphological evidence was found for an attachment of *Escherichia coli* to the intestines of newborn calves. Subsequently, it was shown that biofilms form in aquatic ecosystems under high-shear conditions (Geesey, 1977). Another major advance occurred when scanning confocal laser microscopy was applied to the study of biofilms (Lawrence, 1991; Andersen, 1998). Major advantages of this technology are that the samples are living and hydrated. Consequently, it was discovered that there is a complex organization to biofilms with mushroom- and tower-like structures enclosed in a matrix with intervening water channels (Potera,et. al., 1998; de Beer, et. al., 1994).

Numerous investigations have shown that biofilms have properties, probably derived from their organization, that are of both biological and clinical importance. It has been found consistently that biofilms are far more resistant to static and cidal agents than planktonic cells (Zheng, et. al., 2002; Walters, et. al., 2003; Stewart, et. al., 2004). This applies to a broad spectrum of pathogens, including *Legionella, Listeria, Mycobacteria, Pseudomonas,* and *Vibrio*. There can be as much as a 100 to 1000 times greater resistance in biofilms compared to their planktonic counterparts (Stewart, et. al., 2004; Nickel, et. al., 1985;

Donlan, et. al., 2002). Treating biofilms of *P. aeruginosa* is of medical importance because of its pathogenesis in cystic fibrosis patients, and this has proved a difficult task (Walters, et. al., 2003; Nickel, et. al., 1985; Donlan, et. al., 2002; Costerton, 1999; Elkins, et. al., 1999). The increased resistance of biofilms to antimicrobials is not the result of mutation or horizontal gene transfer because upon removal of the cells from biofilms and their subsequent growth as planktonic cells, the sensitivity to the agents re-emerges (Stewart, et. al., 2004; Costerton, 2004; Elkins, et. al., 1999). The resistance phenomenon also applies to fungal biofilms and has been documented with *Candida albicans* (Kuhn, et. al., 2002; Hawser, et. al., 1995; Chandra, et. al., 2001). *Candida* is cited as being the fourth most-common cause of blood-stream infections (Banerjee, et. al., 1991), and is also a major clinical problem because of its ability to form biofilms on indwelling medical devices (Hawser, et. al., 1995). Due to the ability of microbial cells to adhere to surfaces, both biotic and abiotic, a major problem has developed in medicine with the colonization of indwelling medical devices and nosocomial infections. As the technology of medicine has advanced, and concomitantly the population of the aged is increasing, there has been increased utilization of indwelling medical devices to reduce the length of hospital stays and subsequently increasing dependence on ambulatory care. In addition, patients of all ages use catheters to deliver medications or fluids and devices for those with heart defects and prostheses for joint replacements, particularly hips and knees. Bacterial or fungal biofilms can form on these medical devices and can lead to infection (Zhang, et. al., 1995; Thomas, et. al., 2004; Spoering, et. al., 2001; Nguyen, et. al., 1995; Banerjee, et. al., 1991; Maki, et. al., 1994). Often, the only recourse is to remove the device and replace it (Raad, et. al., 1998). Due to the consequent increase in treatment costs, leading to a serious economic impact and the number of biofilms related infections that arise every year in both hospital and ambulatory care settings, it is important to understand biofilms better and why methods are needed to assess their formation as well as to test their response to antimicrobials.

Methods to assess biofilm formation

There are various strategies used to study biofilm formation, although they generally employ stationary phase cells. These modes of study used static systems, which are better for the study of early events in biofilm formation, and continuous flow or chemostat systems, which are preferable for the examination of mature biofilms (Merritt, et. al., 2005).

Microtiter plate biofilm assay

A popular static method that has the advantage of high throughput is the microtiter plate biofilm assay. This method is based on its development and modification by Christensen et al. (Christensen, et. al., 1985; Mack et al. 1994), and O'Toole and Kolter (O'Toole, et. al., 1998).

Procedure

1. The bacterial strain is grown in batch culture overnight at the desired temperature to stationary phase in the desired medium, such as minimal medium or Luria-Bertani medium. A 10-mL culture or less will suffice.

2. The cells 1:100 are diluted with the desired growth medium into the wells of a 96-well polystyrene microtiter plate. It is recommended that each strain is assayed in quadruplicate. Some wells should be filled with sterile water and the plate should be covered with a lid and/or parafilm to prevent dehydration.

3. The cells are incubated in the plate at appropriate temperature, typically in 25 to 37°C is used (Merritt, et. al., 2005) and the length of the incubation should vary depending on the strain, medium, and the incubation temperature.

4. After the incubation, the planktonic bacteria should be separated from those cells that have formed the biofilm. First, the plates are inverted and firmly shaken over a wastewater tray. Alternatively, the cells can be washed using a water bottle that gently delivers the liquid. The wells should then be washed three more times, either by immersing them into water washes or gently applying water or using appropriate buffer.

Finally, the plates are inverted and tapped to remove any remaining water or buffer.

5. The wells are air dried for few minutes at room temperature and then 0.1% crystal violet solution, 20 to 200 μL are added to each well. Allow the staining to proceed for 10 to 20 min at room temperature.

6. The wells are washed with water to remove crystal violet that has not adhered to the cells. This is repeated three more times and microtiter plates are air dried at room temperature.

7. 200 μL of 95% ethanol are added to each well and incubated for 10 to 15 min at room temperature to solubilize the crystal violet and mixed with a pipette. Other solvents such as dimethylsulfoxide have been used instead of ethanol (Danhorn, et. al., 2004).

8. Transfer 100 to 150 μL of crystal violet solution to corresponding wells of an optically clear, flat-bottom 96-well plate and measure the absorbance at 500 to 600 nm with a plate reader.

9. Alternatively, the crystal violet stain can be quantified by another means (Raad, et. al., 2000):

   a. 200 μL of 95% ethanol are added twice to each microtiter plate well and mixed.

   b. The ethanol is transferred to a 1.5-mL microcentrifuge tube.

   c. The volume is adjusted to 1 mL with deionized water.

   d. Spectrophotometrically read at 540 nm by using a cuvette.

*An in vitro cell culture method for assessing biofilm formatio*n (Merritt, et. al., 2003)

This is a batch culture method, and batch culture is commonly used in studying biofilms because it is easy to grow the cells in a variety of media and vessels and cost is low (McLean, et. al., 2004). In this method, the biofilm is grown on a coverslip that has been inserted into the culture medium.

Procedure

1. Cells are grown overnight in the appropriate medium. The bacteria can be grown aerobically or anaerobically, depending on the organism and the objective of the study.

2. 2.5 mL of the medium is added to individual sterile culture dishes.

3. To each dish, a sterile 18-mm-diameter glass microscope coverslip is used to cover the dish.

4. The requisite amount of overnight culture is added to the dish and incubated overnight.

5. The coverslips are removed and rinsed briefly with water. These can be viewed by phase-contrast microscopy.

6. For fluorescence microscopy, use the following approach:

    a. The coverslips are removed from the culture and transfered to a dry dish.

    b. 20 μL monoclonal antibody solution is added and incubated for 30 min at room temperature (Shi, et. al., 1998).

    c. 5 μL fluorescein-conjugated goat secondary antibody is added and incubated at room temperature for 10 min.

    d. The coverslips are rinsed with water to remove unbound antibodies and cells and then used immediately for microscopy to determine biofilm development.

A static method for assessment of staphylococcus biofilms

It has been well established that indwelling medical devices form excellent surfaces for biofilm formation (Costerton, 1999). *Staphylococcus* spp. – in particular *Staphylococcus epidermidis* and *Staphylococcus aureus* as well as coagulase-negative staphylococci – can produce a mucoid exopolysaccharide (referred to as slime) that allows them to form biofilms on the surfaces of these devices (Raad, et. al., 2000; Donlan, et. al., 2001; Christensen, et. al., 1982). It is these staphylococci that are frequently involved in troublesome infections because of the difficulty in treating biofilms with antibiotics (Stewart, et. al., 2004; Nickel, et. al., 1985; Donlan, et. al., 2002). Biofilm formation can be assessed by a microtiter plate method (Mathur, et. al., 2006) similar to that described above (O'Toole, et. al., 1998; Christensen, et. al., 1985). Biofilm formation can be assessed by a microtiter plate method (Mathur, et. al., 2006) similar to that described above (O'Toole, 1998; Christensen, et. al., 1985). It was found that adding 1% glucose to trypticase soy broth or 2% sucrose to brain heart infusion broth markedly increase the number of strains that formed biofilms (Mathur, et. al., 2006).

As an alternative, a method that requires simpler equipment and detects the slime production is the tube method (Christensen, et. al., 1982; Mathur, et. al., 2006). This is considered a qualitative method, but Mathur et al. (Mathur, et. al., 2006) have shown that it is equivalent to the plate method, at least for strong biofilm formers.

Tube Method Procedure

1. The *Staphylococcus* species is grown on culture plate overnight.
2. 10mL TSB is inoculated containing 1% glucose with a loopful of cells from the culture plate and incubated statically for 18 to 24 hr at 37°C in a glass or polystyrene test tube.
3. The tube is decanted and washed with phosphate buffered saline (pH 7.3).
4. Air dried for up to 10 min.
5. 0.1% crystal violet stain was used for staining 5 to 10 min at room temperature.
6. Washed with deionized water to remove the excess stain.
7. The tubes are inverted and allow to air dry.
8. In assessing the results, the formation of a ring at the liquid interface is not considered a positive result and a visible film that lines the wall and

the bottom of the tube is considered positive (Christensen, et. al., 1982; Mathur, et. al., 2006).

Flow cell method

Microscopic studies have been an important tool in the study of biofilm development. A route to the microscopic examination of developing biofilms is via the flow cell technique. This method was developed to allow direct, on-line observation of biofilm structure by light microscopy and scanning confocal laser microscopy (Wolfaardt, et. al., 1994; McLean, et. al., 2004; Christensen, et. al., 1999). Initially, light microscopy can be used, but as the biofilm thickens, SCLM (scanning confocal laser microscopy) is required to obtain interpretable images.

The scanning consists of five units. A container of medium is connected to a flow pump that is connected to a bubble trap; the medium then flows into a flow cell (chamber), and the effluent then goes to a waste bottle (Wolfaardt, et. al., 1994; Christensen, et. al., 1999). The flow cell has four parallel flow channels (for details of its construction, see Wolfaardt et al. 1994 or Christensen et al. 1999). The bubble chamber is present to prevent air bubbles from reaching the biofilm and disrupting its structure. The tubing between the bubble chamber and the flow cell must be of sufficient length to allow the flow cell to be mounted on a microscope. A critical component of the apparatus is a microscope glass coverslip that serves as a substratum for biofilm development as well as for microscopic observation. According to Christensen et al. (Christensen et al. 1999), the coverslip is turned upward to prevent the sedimentation of flocs detached from other parts of the biofilm.

Procedure

1. The flow cell is sterilized with 0.5% (vol/vol) hypochlorite overnight and then rinsed with distilled water.

2. Using a syringe, 250 μL of an overnight culture is diluted to an A_{450} of 0.05 and is injected into the flow cell while the pump is turned off.

3. The flow cell is turned upside-down.

4. The influent line is clamped to prevent back-growth.

5. After 1 hr, the flow chamber is restored to an upright position and medium is pumped through the system at a rate of 0.2mm/sec.

6. The development of the biofilm can be followed by microscopy.

Modified Robbins device

A motivation for the development of the Robbins device was derived from the problem of biofouling in industry (McCoy, et. al., 1982; McCoy, et. al., 1981; Costerton, et. al., 1984). This instrument, to a large degree, has been supplanted by the modified Robbins device (Nickel, et. al., 1985; McLean, et. al., 2004). With the modified device, materials such as catheter surfaces can be employed to monitor the growth of biofilms (Nickel, et. al., 1985). Consequently, depending on the material chosen, the device can be used for industrial purposes, or for clinical or laboratory investigations (Nickel, et. al., 1985; McLean, et. al., 2004; Domingue, et. al., 1994). It has also been used for the determination of the cidal effects of antibiotics on biofilms (Nickel, et. al., 1985). A key aspect in the design of this instrument is that the sampling ports have disks of the test surface attached at the bottom and in contact with the medium, but the flow of the medium does not disturb the biofilm. Because this is a multiport system, samples can be removed at various times or in multiples. The biofilms that grow can be examined by microscopy or cells can be scraped off the surface of the disk material for enumeration (Nickel, et. al., 1985). The modified Robbins device can be used with continuous cultures systems as well as with batch culture (Millar, et. al., 2001).

Procedure

The experimental design is that of Nickel et al. (1985) which is the seminal publication with respect to the modified Robbins device. In this work, a batch culture was used. The disks attached to the sampling ports were derived from a urinary catheter and the system was organized with a flask that contained the medium fitted with a line containing a valve going to a peristaltic pump. From the pump, medium flowed through tubing into the modified Robbins device and the effluent was collected in an appropriate vessel.

1. The entire experimental apparatus was sterilized using ethylene oxide.

2. Cells *(Pseudomonas aeruginosa)* are added to the medium so that they would be in log phase for the duration of the experiment.

3. The pump rate is maintained at 60 mL hr^{-1}.

4. During the course of the experiment, up to 8 hr, the disks were aseptically removed and examined by scanning electron microscopy or cells are scraped off the disks into medium for enumeration.

Microscopy

Microscopy has been an invaluable tool in studying the development of biofilms (McLean, et. al., 2004; Marrie, 1982; Kinner, et. al., 1985; Costerton, et. al., 2004). Two major modes of microscopy used are scanning electron microscopy (SEM) and scanning confocal laser microscopy (SCLM). The major advantage of SEM is its high resolution; but because of fixation and dehydration of the samples in the preparative phases, it gives a distorted image of biofilm structure. In contrast, SCLM yields a lower resolution; but because the samples are more representative in their natural state, the image obtained is more accurate and has allowed for a much-improved understanding of biofilm structure (Costerton, et. al., 2004).

Preparation of Samples for Scanning Electron Microscopy (Tomaras, et. al., 2003; Marrie, 1984; Marrie, 1983; Ladd, et. al., 1985; Kumon, et. al., 1995)

1. The sample is placed in 2.5 to 5% glutaraldehyde in 0.05 to 0.1 M cacodylate buffer (pH 6.2 to 7.0) for up to 24 hr at 20°C. The mixture often contains 0.15% ruthenium red to stain the exopolysaccharide matrix.

2. If desired, the sample can be washed 2 to 3 times with the buffer for 1 hr each.

3. The sample is treated with 1% osmium tetroxide (Malick, et. al., 1975; Kumon, et. al., 1983).

4. Sample is dehydrated in a series of ethanol wash from 25 to 100%. Each ethanol wash is for 10 min at room temperature.

5. The samples are critical-point dried.

6. The samples are coated with gold or platinum-palladium.

A variation on this procedure, which was used for the study of *Acinetobacter baumanii,* was published more recently by Tomaras et al. (2003). In this study, the bacteria were grown in Petri dishes without shaking.

1. After the appropriate time period, the dishes are flooded with 2.5% glutaraldehyde in 0.05 M cacodylate buffer.

2. This is incubated at room temperature for 2 hr.

3. The glutaraldehyde is removed and the cells are washed with distilled water.

4. Samples of 1 x 1 cm are cut from the Petridish and rinsed gently with distilled water.

5. The preparation is treated with series of ethanol washes as above.

6. The samples are then CO_2-critical-point dried for 60 to 120 min.

7. The samples are then gold-coated.

Scanning Confocal Laser Microscopy (SCLM) (UW Keck Microscopy Facility website; Cowan, et. al., 2000; Jefferson, et. al., 2005; Kuehn, et. al., 1998; Lawrence, 1999)
With SCLM it is possible to view live samples that have not been distorted by fixation. This has allowed for a much better understanding of the three-dimensional organization of biofilms. In addition, SCLM has been an effective experimental tool in conjunction with *in situ* hybridization (Amann, et. al., 1996; Miller, 1996).

Sample Preparation and Analysis

Treatment of the sample in order to view it using SCLM depends on the type of instrument used:

1. Upright microscope: With this type of microscope, the samples are mounted with a material such as acid-free silicone glue.

    a. Biofilm is grown overnight in appropriate medium on a 35-mm, collagen-coated glass coverslip that is sealed to the bottom of a culture dish (Jefferson, et. al., 2005). An alternative is to use a poly-L-lysine-coated coverslip (Cowan, et. al., 2000) glued to an appropriate vessel.

    b. The medium is removed and washed once with water or 0.85% saline.

    c. The coverslip is mounted.

2. Inverted microscope:

    a. Biofilm is grown on an appropriate substratum. Flat surfaces are better because they are easily mounted and stained. Flow cells and the disks of the modified Robbins device are well suited for this.

    b. The biofilm is washed in 0.85% saline or the medium of growth.

    c. The use of an inverted microscope dictates that the growth chamber is mounted upside-down using a coverslip (sealed around with nail polish) at the bottom (UW Keck Microscopy Facility website). This way, the laser light penetrates the coverslip from the bottom of the biofilm to the top, producing an inverted image (Jefferson, et. al., 2005).

3. Staining:

    a. There are a variety of fluorophores for staining nucleic acids. Positive stains include SYTO dyes, acridine orange, DAPI and propidium iodide (Lawrence, et. al., 1999). Fluorescein can be used for either negative or positive staining, depending on the pH (Wolfaardt, et. al., 1994). Depending on the experimental system, the number of fluorophores added must be empirically determined.

b. When more than one fluorescent stain is used, factors such as relative intensity emission characteristics, photostability, interference, additive effects and quenching effects must be considered (Lawrence, et. al., 1999).

4. The mountant: The primary purposes served by the mountant are to preserve the sample and prevent photobleaching. Dimethylsulfoxide (DMSO) is often used with bacterial samples and mixed with the fluorophores before its addition to the sample.

   a. The mountant should have a pH that is compatible with the optimum pH for the fluorophore.

   b. The mountant must have an additive such as DMSO to prevent photobleaching.

   c. The refractive index of the mountant should be as close to that of the sample as possible. For bacteria, this is 1.517 (UW Keck Microscopy Facility website).

5. Magnification: This is much less than with SEM and is typically 40X or 100X in oil immersion. Another option is a 63X water immersion objective lens (Jefferson, et. al., 2005). The sample (with the fluorophores) is then excited with light of appropriate wavelength, causing it to emit light (fluorescence).

6. Imaging:

   a. After locating suitable fields (usually with phase contrast or epi-fluorescence), the appropriate excitation and emission filters are set before scanning the sample.

   b. Either a single thin section (xy plane) is scanned or a series of xy optically thin sections through the biofilm are scanned. This is called a z series or section. It is also referred to as sagittal imaging (Wolfaardt, et. al., 1994). The distance between the z sections depends on the thickness of the biofilm. For example, if the biofilm is approximately 30 to 40 μ thick, then z-slices are obtained every 2 μ (Jefferson, et. al., 2005). Another option is to collect a single or a series of xz vertical sections through the specimen (Lawrence, et. al., 1999).

c. These two-dimensional (2D) cross-sectional images are stacked using the SCLM software (provided by the manufacturer) to get 3D information. With the quantitative extraction of information contained in each image, a three-dimensional reconstruction of the principal biological events can be achieved (Kuehn, et. al., 1998).

7. Image processing and analysis:

a. The SCLM images may require basic processing and enhancement before analysis. These include histogram analysis, gray-level transformation, normalization, contrast enhancement, application of median or lowpass filters or image addition, or multiplication. These are done using specialized digital image processing and analysis software. The processing applied also depends on the information desired. For example, when two different fluorophores are used (e.g., green and red), images of the biofilms can be converted to Adobe Photoshop files (Kuehn, et. al., 1998). A region near the center of the biofilm is selected, a histogram of the region obtained, and the mean intensity value of the green pixels is divided by the mean intensity of the red pixels to obtain a semi-quantitative measure of the accumulation of the fluorophore in the biofilm. Ratios of green/red intensity values obtained for the same region at different time points can be used to calculate the rate of fluorophore transport.

b. Alternatively, images can be generated interactively by calling appropriate command-line scripts using the dialog and menu facilities of the SCLM software (provided by the manufacturer). Then digital image processing and analysis are performed with analytical packages such as the QUANTIMET 570 (Leica, Cambridge, United Kingdom) computer system, which enhances images, discards unwanted details, and accelerates gray-scale processing (Lawrence, et. al., 1999).

c. Computerized image analysis comprises a sequence of operations that depend on the specimen being studied. Before the automated image processing is activated, pixel calibration is done through the QUANTIMET program. The size and position of the area of interest for measurement are specified, and the computer is then programmed to perform a sequence of automated operations such as image acquisition, gray image analysis, detection, measurements of signal intensity, volume integration, and data recording and display (Kuehn, et. al., 1998; Lawrence, et. al., 1999).

d. Deconvolution can also be applied to SCLM images to sharpen the image by mathematical removal of "out-of-focus" information. All these functions help in smoothing the image, reducing the noise, and sharpening the image, thus defining the objects more accurately.

8. Image output: After obtaining the images, they are stored as TIFF, GIFF, PICT, or JPEG, depending on the degree of image fidelity and the degree of image compression required.

Atomic Force Microscopy (AFM)

Atomic force microscopy is an advanced technology for observing the force interactions between its tip and the surface (Binnig, et. al., 1986; Kasemo, et. al., 2002). An atomic force microscope has a nano probe or tip fixed on a sensitive cantilever which can handle the micrometer scale. When the probe is very close to the sample, the top surface of the probe on the cantilever will bend from the original location. A photo detector receives the tiny signal from a laser beam deflected from the cantilever, and transfers/converts it into images viewed in a computer. In AFM, the force between the sample and the tip is used to sense the proximity of the tip to the sample. AFM offers considerable advantages in many scientific areas, including mapping of biological samples such as bacterial biofilms. Chemical modification of the AFM tip by using functional groups such as hydrophilic, hydrophobic, alkyl or carboxyl groups can detect microbial surface functional groups, such as surface hydrophobicity or surface charge.

Recently, electrochemical experimental techniques have been introduced to monitor the growth of bacteria in the process of biofilm growth (such as biofilm age, cell density, pH) as well as macromolecules (extracellular polymeric substances (EPS), such as polysaccharides, protein, humic substances and nucleic acids, as discussed below) produced by the biofilms. Biofilms on the electrode surface have been studied using cyclic voltammetry (CV) (Lower, et. al., 2000), impedance spectroscopy (Hu, et. al., 2007), and infrared reflection spectroscopy (Vostiar, et. al., 2004). Electrochemistry has been used to control biofilm growth, such as stimulation on positively charged solid surfaces and inhibition on negatively charged surfaces (van der Mei, et. al., 2001). Some studies have shown that the negatively charged bacterial biofilms are more likely to grow faster on positively charged electrode surfaces, whereas, negatively charged surfaces inhibit bacterial growth (Bos, et. al., 1999). Other research has shown that static electric fields play a major role in the adhesion of bacterial biofilms to solid surfaces (Kasemo, et. al., 2002; Marsili, et. al., 2008; Timur, et. al., 2007).

The combination of AFM and electrochemistry is a new approach to track and observe the growth of live bacterial biofilms in their natural complex environment with high resolution. Advanced techniques such as AFM and electrochemical methods have been used to carry out investigations of the adhesion mechanisms of biofilms on various substrates, e.g., mica, glass, and pure and self-assembled monolayer (SAM) gold AU (111) surfaces (parentheses indicate the most stable isotope of gold). A SAM, an organized molecular assembly of amphiphilic molecules, is formed by adsorption onto a solid surface from a homogeneous solution. This organization is achieved due to the affinity of the head group for the surface, and from the slow two-dimensional orientations of the hydrophobic tail groups.

Preparation of the sample and the AFM tip

Immobilization of cells

AFM requires the sample to be attached to a solid substrate, which can be a glass slides, mica, or metal surface. Glass is composed of silicon dioxide and is not atomically flat; whereas, mica can provide atomically flat surfaces that are ideal as a supporting substrate for AFM studies. Biosamples can be immobilized on a solid surface by physical adsorption or chemical adsorption. For physical adsorption, a drop of sample solution is put on a surface and dried in air. For chemical adsorption, a functional chemical is used to immobilize the target sample on a surface through formation of chemical bonds. L-lysine, fibronectin and agglutinin are examples of linkers for chemical adsorption or electrostatic adsorption. Along this line, a surface modified by a chemical can change its original physical or chemical properties and increase the adhesion of cells. Denaturation of biological samples is an essential problem for any immobilization method. Much effort has been put into seeking an approach with high AFM resolution and little denaturation. Mechanical fixation of individual cells in a polymer membrane is among the most successful developments. Some approaches with modified surfaces, like SAMs on AU (111) or the change of molecular charge on a surface, offer a promise to improve bacterial adhesion (Busalmen, et. al., 2007). Some modes of attachment on a solid surface use extracellular polymeric substance or agar gel or a porous membrane to immobilize the cells mechanically. In the latter, spherical cells are trapped in a polycarbonate membrane with a pore size similar to that of the cells to obtain high-resolution images without cell attachment or cell damage (Gaboriaud, et. al., 2007). This approach has two advantages: (1), it is simple and does not involve drying, coating or chemical fixation; and (2), it can be used for AFM observations *in situ*, which is close to the natural environment during bacterial growth.

Chemically modified tips

A chemically modified tip is a prerequisite for a reliable, quantitative surface force measurement. The surfaces of commercial tips have defined surface coatings, such as silicon nitride, or gold coated films, which are easily contaminated. Functional tips with SAMs of alkanethiols give high-resolution images due to their chemical sensitivity. The procedure is described in the experimental section (Sub-section describing preparation of hydrophobic tips and surfaces under the procedures section). Alkanethiols with different terminal functional groups: -OH, $-CH_3$, -COOH, $-NH_2$, can be used to detect the physical and chemical properties (hydrophobicity, charge) of biological surfaces.

Biologically modified tips

Some biomolecules are used to modify tips with the purpose of measuring intermolecular forces between individual ligand-receptor contacts, the adhesion of the biofilms or the receptor protein location on single cell surfaces (Ahimou, et. al., 2002). The binding of the biomolecules to the tips should be much stronger than the intermolecular force.

Cell modified probes

A method has been developed to attach cells directly onto an AFM cantilever. This approach can probe cell-solid and cell-cell interactions. This method preserves the native ultrastructure of the cells during measuring. Firstly, cells were immobilized by using glutaraldehyde treatment to create covalent crosslinking between the bacterial layer and the polyethylene glycol, polyethylenimine (PEG-PEI)-coated tip, or by attaching a single yeast cell with a small amount of glue (Razatos, et. al., 1998). Lower and coworkers (Benoit, et. al., 2002) adsorbed bacteria onto poly-L-lysine-coated glass beads, and subsequently attached a cell-coated bead to a cantilever by using a small amount of epoxy resin. Another elegant approach is to attach individual cells to an AFM cantilever via lectins such as wheat germ agglutinin (Oh, et. al., 2009). However, these treatments are likely to affect the structure and properties of the cell surface.

General procedure

Cleaning

All glassware and the cells for both electrochemistry and AFM are boiled in 15% nitric acid, and washed thoroughly with Millipore-grade water. The glassware and the cells are then placed in an ultrasound bath for at least 20 minutes, and washed again to ensure that all impurities have been removed. Both the electrochemical cell and the AFM cell were kept in a beaker filled with Millipore-grade water, and sealed with parafilm after cleaning.

Procedure for cleaning gold surfaces

All gold electrodes were annealed in an oven at 860°C overnight. Prior to use, the surfaces were flame-annealed in a hydrogen flame, and quenched in Millipore-grade water saturated with hydrogen gas. Gold surfaces that had been in contact with metal ions electro polished in 3M sulfuric acid before the annealing procedures. The electro-oxidation produces a thin crust of gold oxide containing possible surface impurities, which are subsequently removed by dissolution in concentrated (36-38%) hydrochloric acid.

Self-assembled monolayers (SAM)

Self-assembled monolayers on Au (111) surfaces were prepared by immersion of the freshly annealed metal surfaces into a 1-2mM solution of the monolayer-forming compound for 24 hours. The solvent was either Millipore-grade water or ethanol as required. The surfaces were thoroughly rinsed in both solvent and Millipore-grade water before further use.

Procedure

1. Biofilm growth

    a. Mica is used for growing biofilms. 2.0x1.0 cm mica sheets are cut and freshly cleaved. Prior to use the sheets are sterilized by immersion into sodium hypochlorite solution (0.5%) for 2h.

    b. The mica slices are washed three times with sterilized Millipore-grade water, autoclaved at 121°C and 15 psi for 30 min and immersed into a Petridish containing 10 ml of nutrient medium. Typically, bacteria are cultivated on mica in Trypticase Soy Broth (TSB) for 4 hours and 24 hours.

    c. Overnight cultures grown in appropriate medium are diluted to OD_{600} = 0.05 into fresh TSB medium, added over the mica, sealed by parafilm and incubated for appropriate period at 37°C.

The mica is then washed three times with Millipore-grade water and put into a new Petridish to dehydrate naturally for one day.

2. Preparation of hydrophobic tips and surfaces

   Gold coated surfaces are obtained as follow:

   a. Silicon wafers are coated with electron beam thermal evaporation with a 5 nm thick chromium layer followed by a 30 nm thick gold Au (111) layer.

   b. The Au (111) substrate is coated with a SAM of octadecanethiol. The gold coated surfaces and gold-coated AFM tips are cleaned for 15 min by UV and ozone treatment, rinsed with ethanol, dried with a gentle nitrogen flow, and immersed overnight in a 1 mM solution of octadecanethiol in ethanol and then rinsed in ethanol.

3. AFM measurements and image acquisition.

   a. AFM measurements in Millipore-grade water are performed using a Nanoscope 111a AFM and oxide sharpened microfabricated silicon nitride cantilevers. To probe the microbes in their native state by AFM, the biofilm bacteria are immobilized by mechanical trapping into the mica surface. The mounted sample is transferred into the AFM liquid cell while avoiding dewetting, which can occur at a solid-liquid or liquid-liquid interface. Generally, dewetting describes the rupture of a thin liquid film on the substrate (either a liquid itself or a solid) and the formation of droplets. The imaging force is kept as low as possible (=250pN) to minimize sample damage.

   b. To study adhesion properties of biofilms, both topographical and deflection images were recorded as they yield complementary information. Topographical images provide quantitative information on sample surface topography (surface roughness and height measurements). Deflection images often exhibit a higher contrast of the morphological details, and are especially useful for rough surfaces. Real time images are collected at time intervals, for example every 15 minutes.

c. To study biofilm structure and dynamics in a natural aqueous medium, and other aspects of the chemical as well as cell surface properties important in biofilm formation and growth, a DualScope microscope equipped with a C-21 controller can be used. The tapping mode is used to record all images. A second instrument that can be employed is a PicoScan 5500 with a 100 µm scanner and cantilevers (NP-S). This instrument is used in the contact mode.

Methods for examining the effect of antibiotics or biocides on biofilms

It is now accepted that bacterial growth in natural settings is most commonly in the form of biofilms. From a clinical perspective this presents a problem because biofilms grow on indwelling medical devices, particularly central venous catheters (Dickinson, 1989), and may be involved in the persistence of chronic infections (Christensen, et. al., 1999). A common experience is that the biofilms are considerably more resistant to antibiotics and biocides than planktonic cells (Ceri, et. al., 2001; Ceri, et. al., 1999; Nickel, et. al., 1985). Consequently, methods have been developed to examine the inhibition or killing of biofilms by antimicrobials, which have led to the term "minimal biofilm eradication concentrations" (MBECs) (Ceri, et. al., 2001; Ceri, et. al., 1999). These include the use of the modified Robbins device, the Calgary Biofilm Device, and simpler methods.

The Modified Robbins Device (Nickel, et. al., 1985; Raad, et. al., 1995)

As detailed above, the modified Robbins device was designed primarily to measure biofilm formation. However, it can be used to determine the ability of antibiotics to inhibit the growth or kill cells in a biofilm (Nickel, et. al., 1985; Raad, et. al., 1995).

Procedure

The cells are grown in the modified Robbins device to form a biofilm as described above.

1. After the biofilm has formed, the valve of a parallel reservoir is opened containing antibiotic in the same medium.

2. After the appropriate period of antibiotic exposure, aseptically scrape the cells on a disk (urinary catheter) from a specimen plug into a tube containing sterile phosphate-buffered saline.

3. Subject the cells along with the disk to low-output sonication to disperse the cells.

4. Dilute the cells as needed and plate on nutrient agar for plate counting.

The Calgary Biofilm Device (Bardouniotis, et. al., 2001; Ceri, et. al., 2001; Ceri, et. al., 1999; McLean, et. al., 2004)

Although modified Robbins device could be used to determine the susceptibility of biofilms to antibiotics, it was deemed less than optimal for rapid antibiotic susceptibility testing for clinical purposes (Ceri, et. al., 1999). The Calgary biofilm device is relatively simple in that it consists of two parts that are composed of plastic. One is the bottom, which contains 96 troughs (channels) for bacterial growth, and the other is a lid into which pegs are placed to fit into the troughs. The biofilms are grown on the pegs. The pegs will also fit into standard 96-well polystyrene microtiter plates (Ceri, et. al., 2001). This group demonstrated that the device can produce 96 equivalent biofilms and is highly reproducible. Advantages of this system are that it requires no pumps or tubing thereby reducing problems of contamination, allows for multiple testing, has a platform that uses standard 96-well plates and is considerably faster than the modified Robbins device (Ceri et al., 2001; Ceri, et. al., 1999). This system is also referred to as the MBEC (minimal biofilm eradication concentration) Assay System.

Procedure

1. Establish biofilm conditions:

    a. A standardized inoculum is added to the channels in the bottom plate using TSB as the growth medium.

    b. The lid is attached and placed on a rocking table. The rocking action creates shear forces that promote biofilm formation. For aerobic cultures, the temperature is usually 35°C with 95% relative humidity.

    c. At selected times, duplicate pegs are removed and placed in 200 μL sterile buffer and sonicated for 5 min using a sonic cleaning system.

    d. Viable counts are determined by plating on an appropriate medium.

2. Susceptibility testing:

    a. After generating a standardized biofilm, susceptibility to antibiotics or biocides can be tested.

    b. Working solutions of antibiotics may be prepared in cation-adjusted Mueller-Hinton broth (CAMHB) to a concentration of 1024 μg mL^{-1}.

    c. The antibiotics are diluted in CAMHB and added to the wells of a 96-well plate.

    d. Wash the pegs (in the lid) on which biofilms have grown in a wash tray containing sterile buffer.

    e. Place the lid (with the pegs) over the wells to expose the pegs with the biofilms to the antibiotics. The assembly is usually incubated overnight.

    f. The MBEC value is determined (the minimal concentration of antibiotic that prevents growth of the biofilm).

3. The lid with the pegs is washed as above.

4. The lid is then placed over a 96-well plate in which the wells contain recovery medium.

5. Sonication as above is performed to disperse the cells into the recovery medium.

6. Viable cell number is determined by plating onto the medium of choice.

7. Alternatively, turbidity is measured in wells using a 96-well plate reader.

Other methods for assessing the effects of antibiotics or biocides on biofilm viability (Anderl, et. al., 2000; Elkins, et. al., 1999; Stone, et. al., 2002)

Method 1

Stone et al. (2002) examined the effect of tetracycline on biofilms of uropathogenic *Escherichia coli* grown in Plexiglas flow cells.

1. For enumeration, biofilm is aseptically scraped into 400 μL of 0.1M phosphate buffer at pH 7.0.

2. The cells are vigorously vortexed.

3. The suspensions are diluted and plated on LB agar for enumeration.

Method 2

Anderl et al. (2000) examined the effect of ampicillin and ciprofloxacin on *Klebsiella pneumonia.* In this study the biofilms were propagated on polycarbonate filters.

1. The biofilm is enumerated by immersing the membrane in 9 mL phosphate-buffered water.

2. The membrane is vigorously vortexed to suspend the cells.

3. Serial dilutions on R2A agar is plated by the drop plating method (Hoben, et. al., 1982).

Method 3

Elkins et al. (Elkins, et. al., 1999) studied the killing of planktonic cells and biofilms of *Pseudomonas aeruginosa* by hydrogen peroxide. The biofilms were generated using a drip-flow reactor with stainless steel slides as the substratum.

1. To enumerate the biofilms, slides are removed and scraped aseptically into 50 mL phosphate-buffered saline (pH 7.2) containing 0.2% sodium thiosulfate to neutralize the hydrogen peroxide.

2. To suspend the cells, the above mixture is homogenized in a PT 10/35 Brinkman homogenizer for 15 sec at setting 4.

3. The homogenate is appropriately diluted and plated on R2A agar medium.

Genetic methods

A. *Polymerase Chain Reaction* (PCR)

Based on PCR, various analytical techniques can be used in order to identify biofilm organisms without prior cultivation.

1. DNA fragments have to be extracted from the sample (Ogram, et. al., 1987). It may be suitable to disrupt the organisms, e.g., by ultrasonic treatment, then they are concentrated by membrane filtration and lysed afterward (biofilms require mechanical lysis followed by chemical lysis to disrupt the membranes for complete extraction of DNA). It is also possible to lyse the cells directly in the sample, and obtain the DNA by high speed centrifugation. The use of polyvinylpolypyrrolidone is recommended in order to remove humic substances if these are present in the sample as they can interfere with the PCR reaction.

2. After amplification of the extracted rDNA, a mixture of the amplified products is obtained which has to be separated by molecular biological techniques and subsequently cloned.

3. Restriction analysis of the cloned rDNA fragments is performed and the pattern of the fragment analysis can be compared to that of the culturable organisms of the population. In this way it is possible to obtain genetic

fingerprints of bacteria, protozoa, fungi, possibly present in the sample. An efficient analysis of the microorganisms by means of PCR fingerprinting can be supported by using DNA sequencing, capillary electrophoresis, and databases with reference patterns for the identification of unknown isolates.

B. Gene Probes (Holben, et. al., 1988; Ogram, et. al., 1988)
With the use of rDNA directed and fluorescence labeled oligonucleotide probes, it is possible to detect and localize microorganisms in a biofilm which may not be culturable and identify them phylogenetically. More than one oligonucleotide probe can be applied in order to label more than one kind of organism. This method provides not only information about the species present in the undisturbed sample, but also reveals their position in relation to each other, using the potential of SCLM to demonstrate three dimensional structures. This approach makes it particularly suitable to investigate the functional architecture of a biofilm and the intensity of the signal depends on upon the ribosome content and thus reflects the physiological status.

Though the gene probe strategy has brought significant progress to biofilm research, a drawback is that in order to apply the probes, the cells must be permeabilized. The procedure required for this purpose may kill them, and therefore the possibility of affecting the architecture of the biofilm cannot be completely excluded.

Immunological Methods

Detection and identification of biofilm population members can also be performed by immunological coupling reactions between fluorescent labeled antibodies which are specific to certain organisms. This method requires cultivation of the target organisms, and the availability of antibodies that can detect suitable epitopes in a thin biofilm. In multilayered biofilms, the EPS matrix acts as a diffusion barrier for large molecules and the antibodies will be impeded in reaching the antigenic sites of the target organisms. The advantage in the use of mono-or polyclonal antibodies is that the cells are not killed by this method and thus cellular processes in the biofilm still can be observed. Also the combination with other non-destructive fluorescence staining methods is possible, e.g., using the tetrazolium salt 5-cyano-2,3-ditolyltetrazolium chloride.

Conclusions and future trends

A variety of techniques including conventional and classical techniques, microscopy, genetic and immunological methods are used for assessments are available or are further being developed for assessment of biofilm structure, development, composition, diversity of microbes that are present in biofilms. Many novel techniques are also further developed. A combination of these techniques could elucidate insights about biofilms that could aid research and applications in the field of biofilms.

References

Ahimou, F., Denis, F.A., Touhami, A., and Dufrêne, Y.F. (2002). Probing Microbial Cell Surface Charges by Atomic Force Microscopy. Langmuir *18*, 9937-9941.

Alsteens, D., Dague, E., Rouxhet, P.G., Baulard, A.R., and Dufrêne, Y.F. (2007). Direct Measurement of Hydrophobic Forces on Cell Surfaces Using AFM. Langmuir *23*, 11977-11979.

Amann, R., Snaidr, J., Wagner, M., Ludwig, W., and Schleifer, K.H. (1996). In situ visualization of high genetic diversity in a natural microbial community. Journal of Bacteriology *178*, 3496-3500.

Anderl, J.N., Franklin, M.J., and Stewart, P.S. (2000). Role of Antibiotic Penetration Limitation in Klebsiella pneumoniae Biofilm Resistance to Ampicillin and Ciprofloxacin. Antimicrobial Agents and Chemotherapy *44*, 1818-1824.

Andersen, J.B., Sternberg, C., Poulsen, L.K., Bjorn, S.P., Givskov, M., and Molin, S. (1998). New unstable variants of green fluorescent protein for studies of transient gene expression in bacteria. Applied and environmental microbiology *64*, 2240-2246.

Banerjee, S.N., Emori, T.G., Culver, D.H., Gaynes, R.P., Jarvis, W.R., Horan, T., Edwards, J.R., Tolson, J., Henderson, T., and Martone, W.J. (1991). Secular trends in nosocomial primary bloodstream infections in the United States, 1980–1989. The American Journal of Medicine *91*, S86-S89.

Bardouniotis, E. (2001). Characterization of biofilm growth and biocide susceptibility testing of Mycobacterium phlei using the MBEC™ assay system. FEMS Microbiology Letters *203*, 263-267.

Beech, I.B., Smith, J.R., Steele, A.A., Penegar, I., and Campbell, S.A. (2002). The use of atomic force microscopy for studying interactions of bacterial biofilms with surfaces. Colloids and Surfaces B: Biointerfaces *23*, 231-247.

Benoit, M. (2002). Cell Adhesion Measured by Force Spectroscopy on Living Cells. In Atomic Force Microscopy in Cell Biology (Elsevier), pp. 91-114.

Binnig, G., Quate, C.F., and Gerber, C. (1986). Atomic Force Microscopy. Physical Review Letters *56*, 930-933.

Bogdanov, A., Simonova, M., and Weissleder, R. (1998). Design of metal-binding green fluorescent protein variants. Biochimica et Biophysica Acta (BBA) - Gene Structure and Expression *1397*, 56-64.

Bos, R. (1999). Physico-chemistry of initial microbial adhesive interactions – its mechanisms and methods for study. FEMS Microbiology Reviews *23*, 179-229.

Busalmen, J.P., Berná, A., and Feliu, J.M. (2007). Spectroelectrochemical Examination of the Interaction between Bacterial Cells and Gold Electrodes. Langmuir *23*, 6459-6466.

Butterfield, C.T. (1929). Experimental Studies of Natural Purification in Polluted Waters: III. A Note on the Relation between Food Concentration in Liquid Media and Bacterial Growth. Public Health Reports (1896-1970) *44*, 2865.

Ceri, H., Olson, M., Morck, D., Storey, D., Read, R., Buret, A., and Olson, B. (2001). [25] The MBEC assay system: Multiple equivalent biofilms for antibiotic and biocide susceptibility testing. In Microbial Growth in Biofilms - Part B: Special Environments and Physicochemical Aspects (Elsevier), pp. 377-385.

Ceri, H., Olson, M.E., Stremick, C., Read, R.R., Morck, D., and Buret, A. (1999). The Calgary Biofilm Device: new technology for rapid determination of antibiotic susceptibilities of bacterial biofilms. Journal of clinical microbiology *37*, 1771-1776.

Ceri, H., Storey, D., Morck, D., and Olson, M. (2005). Minimal Biofilm Eradication Concentration (MBEC) Assay. In Biofilms, Infection, and Antimicrobial Therapy (CRC Press), pp. 257-269.

Chandra, J., Mukherjee, P.K., Leidich, S.D., Faddoul, F.F., Hoyer, L.L., Douglas, L.J., and Ghannoum, M.A. (2001). Antifungal Resistance of Candidal Biofilms Formed on Denture Acrylic in vitro. Journal of Dental Research *80*, 903-908.

Christensen, B.B., Sternberg, C., Andersen, J.B., Palmer, R.J., Toftgaard Nielsen, A., Givskov, M., and Molin, S. (1999). [2] Molecular tools for study of biofilm physiology. In Biofilms (Elsevier), pp. 20-42.

Christensen, G.D. (1982). Nosocomial Septicemia Due to Multiply Antibiotic-Resistant Staphylococcus epidermidis. Annals of Internal Medicine *96*, 1.

Christensen, G.D., Simpson, W.A., Younger, J.J., Baddour, L.M., Barrett, F.F., Melton, D.M., and Beachey, E.H. (1985). Adherence of coagulase-negative staphylococci to plastic tissue culture plates: a quantitative model for the adherence of staphylococci to medical devices. Journal of clinical microbiology *22*, 996-1006.

Costerton, J.W. (1999). Bacterial Biofilms: A Common Cause of Persistent Infections. Science *284*, 1318-1322.

Costerton, J.W., Cheng, K.J., Geesey, G.G., Ladd, T.I., Nickel, J.C., Dasgupta, M., and Marrie, T.J. (1987). Bacterial Biofilms in Nature and Disease. Annual Review of Microbiology *41*, 435-464.

Costerton, J.W., Ghannoum, M., Toole, G.A., Washington, D.C., and M., A.S. (2004). A short history of the development of the biofilm concept, in Microbial Biofilms, and O' Eds., chap. 1. *2004 SRC - BaiduScholar*.

Costerton, J.W., and Lashen, E.S. (1984). Influence of biofilm on efficacy of biocides on corrosion-causing bacteria, Mater. Perform., 34. *23 SRC - BaiduScholar*.

Cowan, S.E., Gilbert, E., Khlebnikov, A., and Keasling, J.D. (2000). Dual Labeling with Green Fluorescent Proteins for Confocal Microscopy. Applied and Environmental Microbiology *66*, 413-418.

Danhorn, T., Hentzer, M., Givskov, M., Parsek, M.R., and Fuqua, C. (2004). Phosphorus Limitation Enhances Biofilm Formation of the Plant Pathogen Agrobacterium tumefaciens through the PhoR-PhoB Regulatory System. Journal of Bacteriology *186*, 4492-4501.

de Beer, D., Stoodley, P., Roe, F., and Lewandowski, Z. (1994). Effects of biofilm structures on oxygen distribution and mass transport. Biotechnology and Bioengineering *43*, 1131-1138.

Dickinson, G.M., and Bisno, A.L. (1989). Infections associated with indwelling devices: concepts of pathogenesis; infections associated with intravascular devices. Antimicrobial Agents and Chemotherapy *33*, 597-601.

Domingue, G., Ellis, B., Dasgupta, M., and Costerton, J.W. (1994). Testing antimicrobial susceptibilities of adherent bacteria by a method that incorporates guidelines of the National Committee for Clinical Laboratory Standards. Journal of clinical microbiology *32*, 2564-2568.

Donlan, R. (2001). Biofilms and Device-Associated Infections. Emerging Infectious Diseases *7*, 277-281.

Donlan, R.M., and Costerton, J.W. (2002). Biofilms: Survival Mechanisms of Clinically Relevant Microorganisms. Clinical Microbiology Reviews *15*, 167-193.

Elkins, J.G., Hassett, D.J., Stewart, P.S., Schweizer, H.P., and McDermott, T.R. (1999). Protective role of catalase in Pseudomonas aeruginosa biofilm resistance to hydrogen peroxide. Applied and environmental microbiology *65*, 4594-4600.

Gaboriaud, F., and Dufrêne, Y.F. (2007). Atomic force microscopy of microbial cells: Application to nanomechanical properties, surface forces and molecular recognition forces. Colloids and Surfaces B: Biointerfaces *54*, 10-19.

Gaylarde, C., and Cook, P. (1990). New rapid methods for the identification of sulphate-reducing bacteria. International Biodeterioration *26*, 337-345.

Geesey, G.G., Richardson, W.T., Yeomans, H.G., Irvin, R.T., and Costerton, J.W. (1977). Microscopic examination of natural sessile bacterial populations from an alpine stream. Canadian Journal of Microbiology *23*, 1733-1736.

Gião, M.S., Montenegro, M.I., and Vieira, M.J. (2003). Monitoring biofilm formation by using cyclic voltammetry - effect of the experimental conditions on biofilm removal and activity. Water Science and Technology *47*, 51-56.

Hassett, D.J., Ma, J.-F., Elkins, J.G., McDermott, T.R., Ochsner, U.A., West, S.E.H., Huang, C.-T., Fredericks, J., Burnett, S., Stewart, P.S., *et al.* (1999). Quorum sensing in Pseudomonas aeruginosa controls expression of catalase and superoxide dismutase genes and mediates biofilm susceptibility to hydrogen peroxide. Molecular Microbiology *34*, 1082-1093.

Hawser, S.P., and Douglas, L.J. (1995). Resistance of Candida albicans biofilms to antifungal agents in vitro. Antimicrobial Agents and Chemotherapy *39*, 2128-2131.

Hoben, H.J., and Somasegaran, P. (1982). Comparison of the Pour, Spread, and Drop Plate Methods for Enumeration of Rhizobium spp. in Inoculants Made from Presterilized Peat. Applied and environmental microbiology *44*, 1246-1247.

Holben, W.E., and Tiedje, J.M. (1988). Applications of Nucleic Acid Hybridization in Microbial Ecology. Ecology *69*, 561-568.

Hu, Y., Zhang, J., and Ulstrup, J. (2010). Interfacial Electrochemical Electron Transfer Processes in Bacterial Biofilm Environments on Au(111). Langmuir *26*, 9094-9103.

Hu, Z., Jin, J., Abruña, H.D., Houston, P.L., Hay, A.G., Ghiorse, W.C., Shuler, M.L., Hidalgo, G., and Lion, L.W. (2007). Spatial Distributions of Copper in Microbial Biofilms by Scanning Electrochemical Microscopy. Environmental Science & Technology *41*, 936-941.

Jefferson, K.K., Goldmann, D.A., and Pier, G.B. (2005). Use of Confocal Microscopy To Analyze the Rate of Vancomycin Penetration through Staphylococcus aureus Biofilms. Antimicrobial Agents and Chemotherapy *49*, 2467-2473.

Kasemo, B. (2002). Biological surface science. Surface Science *500*, 656-677.

Kinner, N.E., Maratea, D., and Bishop, P.L. (1985). An electron microscopic evaluation of bacteria inhabiting rotating biological contactor biofilms during various loading conditions. Environmental Technology Letters *6*, 455-466.

Kuehn, M., Hausner, M., Bungartz, H.J., Wagner, M., Wilderer, P.A., and Wuertz, S. (1998). Automated confocal laser scanning microscopy and semiautomated image processing for analysis of biofilms. Applied and environmental microbiology *64*, 4115-4127.

Kuhn, D.M., George, T., Chandra, J., Mukherjee, P.K., and Ghannoum, M.A. (2002). Antifungal Susceptibility of Candida Biofilms: Unique Efficacy of Amphotericin B Lipid Formulations and Echinocandins. Antimicrobial Agents and Chemotherapy *46*, 1773-1780.

Kumon, H., Kaneshige, T., and Ohmori, H. (1983). T4-labeling technique and its applications with particular reference to blood group antigens in bladder tumors. Scanning electron microscopy, 939-948.

Kumon, H., Ono, N., Iida, M., and Nickel, J.C. (1995). Combination effect of fosfomycin and ofloxacin against Pseudomonas aeruginosa growing in a biofilm. Antimicrobial Agents and Chemotherapy *39*, 1038-1044.

Ladd, T.I., Schmiel, D., Nickel, J.C., and Costerton, J.W. (1985). Rapid method for detection of adherent bacteria on Foley urinary catheters. Journal of clinical microbiology *21*, 1004-1006.

Lawrence, J.R., Korber, D.R., Hoyle, B.D., Costerton, J.W., and Caldwell, D.E. (1991). Optical sectioning of microbial biofilms. Journal of Bacteriology *173*, 6558-6567.

Lawrence, J.R., and Neu, T.R. (1999). Confocal laser scanning microscopy for analysis of microbial biofilms. In Biofilms (Elsevier), pp. 131-144.

Loeb, G.I. (1980). Measurement of microbial marine fouling films by light section microscopy, Mar. Technol. Soc. J., 14. *14 SRC - BaiduScholar.*

Long, E.R. (1972). Studies of marine fouling and boring off Kodiak Island, Alaska. Marine Biology *14*, 52-57.

Lower, S.K., Tadanier, C.J., and Hochella, M.F. (2000). Measuring interfacial and adhesion forces between bacteria and mineral surfaces with biological force microscopy. Geochimica et Cosmochimica Acta *64*, 3133-3139.

Mack, D., Nedelmann, M., Krokotsch, A., Schwarzkopf, A., Heesemann, J., and Laufs, R. (1994). Characterization of transposon mutants of biofilm-producing Staphylococcus epidermidis impaired in the accumulative phase of biofilm production: genetic identification of a hexosamine-containing polysaccharide intercellular adhesin. Infection and immunity *62*, 3244-3253.

Maki, D.G., Bisno, A.L., Waldovogel, F.A., Washington, D.C., and M., A.S. (1994). Infections caused by intravascular devices used for infusion therapy: pathogenesis, prevention, and management, in Infections Associated with Indwelling Medical Devices, 2nd edition, and Eds. p 155 *1994 SRC - BaiduScholar.*

Malick, L.E., Wilson, R.B., and Stetson, D. (1975). Modified Thiocarbohydrazide Procedure for Scanning Electron Microscopy: Routine use for Normal, Pathological, or Experimental Tissues. Stain Technology *50*, 265-269.

Marrie, T.J., and Costerton, J.W. (1983). Scanning electron microscopic study of uropathogen adherence to a plastic surface. Applied and environmental microbiology *45*, 1018-1024.

Marrie, T.J., and Costerton, J.W. (1984). Morphology of bacterial attachment to cardiac pacemaker leads and power packs. Journal of clinical microbiology *19*, 911-914.

Marrie, T.J., Nelligan, J., and Costerton, J.W. (1982). A scanning and transmission electron microscopic study of an infected endocardial pacemaker lead. Circulation *66*, 1339-1341.

Marsili, E., Rollefson, J.B., Baron, D.B., Hozalski, R.M., and Bond, D.R. (2008). Microbial Biofilm Voltammetry: Direct Electrochemical Characterization of Catalytic Electrode-Attached Biofilms. Applied and Environmental Microbiology *74*, 7329-7337.

Mathur, T., Singhal, S., Khan, S., Upadhyay, D.J., Fatma, T., and Rattan, A. (2006). Detection of biofilm formation among the clinical isolates of Staphylococci: An evaluation of three different screening methods. Indian Journal of Medical Microbiology *24*, 25.

McCoy, W.F., Bryers, J.D., Robbins, J., and Costerton, J.W. (1981). Observations of fouling biofilm formation. Canadian Journal of Microbiology *27*, 910-917.

McCoy, W.F., and Costerton, J.W. (1982). Fouling biofilm development in tubular flow systems, Dev. Indust. Microbiol., 551. *23 SRC - BaiduScholar.*

McLean, R.J.C., McGowin, C.L., Barnes, M.B., Aron, G.M., and Bates, C.L. (2004). Methods of Studying Biofilms. In Microbial Biofilms (American Society of Microbiology), pp. 379-413.

Merritt, J., Qi, F., Goodman, S.D., Anderson, M.H., and Shi, W. (2003). Mutation of luxS Affects Biofilm Formation in Streptococcus mutans. Infection and Immunity *71*, 1972-1979.

Merritt, J.H., Kadouri, D.E., and O'Toole, G.A. (2005). Growing and Analyzing Static Biofilms. In Current Protocols in Microbiology (John Wiley & Sons, Inc.).

Millar, M.R., Linton, C.J., and Sherriff, A. (2001). [3] Use of a continuous culture system linked to a modified robbins device or flow cell to study attachment of bacteria to surfaces. In Microbial Growth in Biofilms - Part B: Special Environments and Physicochemical Aspects (Elsevier), pp. 43-62.

Miller, J.A., and Ghanekar, D. (1996). Pneumothoraces secondary to blunt abdominal trauma: aids to plain film radiographic diagnosis and relationship to solid organ injury. The American surgeon *62*, 416-420.

Nedelmann, M., Sabottke, A., Laufs, R., and Mack, D. (1998). Generalized Transduction for Genetic Linkage Analysis and Transfer of Transposon Insertions in Different Staphylococcus epidermidis Strains. Zentralblatt für Bakteriologie *287*, 85-92.

Nguyen, M.H. (1995). Therapeutic Approaches in Patients With Candidemia. Archives of Internal Medicine *155*, 2429.

Nickel, J.C., Ruseska, I., Wright, J.B., and Costerton, J.W. (1985). Tobramycin resistance of Pseudomonas aeruginosa cells growing as a biofilm on urinary catheter material. Antimicrobial Agents and Chemotherapy *27*, 619-624.

Ogram, A., Sayler, G.S., and Barkay, T. (1987). The extraction and purification of microbial DNA from sediments. Journal of Microbiological Methods *7*, 57-66.

Ogram, A.V., and Sayler, G.S. (1988). The use of gene probes in the rapid analysis of natural microbial communities. Journal of Industrial Microbiology *3*, 281-292.

Oh, Y.J., Jo, W., Yang, Y., and Park, S. (2007). Influence of culture conditions on Escherichia coli O157:H7 biofilm formation by atomic force microscopy. Ultramicroscopy *107*, 869-874.

Oh, Y.J., Lee, N.R., Jo, W., Jung, W.K., and Lim, J.S. (2009). Effects of substrates on biofilm formation observed by atomic force microscopy. Ultramicroscopy *109*, 874-880.

O'Toole, G.A., and Kolter, R. (1998). Initiation of biofilm formation in Pseudomonas fluorescens WCS365 proceeds via multiple, convergent signalling pathways: a genetic analysis. Molecular Microbiology *28*, 449-461.

Pedersen, A.R., Møller, S., Molin, S., and Arvin, E. (1997). Activity of toluene-degrading Pseudomonas putida in the early growth phase of a biofilm for waste gas treatment. Biotechnology and Bioengineering *54*, 131-141.

Potera, C. (1998). Studying Slime. Environmental Health Perspectives *106*, A604.

Raad, I. (1998). Intravascular-catheter-related infections. The Lancet *351*, 893-898.

Raad, I. (2000). Management of intravascular catheter-related infections. Journal of Antimicrobial Chemotherapy *45*, 267-270.

Raad, I., Darouiche, R., Hachem, R., Sacilowski, M., and Bodey, G.P. (1995). Antibiotics and prevention of microbial colonization of catheters. Antimicrobial Agents and Chemotherapy *39*, 2397-2400.

Razatos, A., Ong, Y.L., Sharma, M.M., and Georgiou, G. (1998). Molecular determinants of bacterial adhesion monitored by atomic force microscopy. Proceedings of the National Academy of Sciences *95*, 11059-11064.

Schadow, K.H., Simpson, W.A., and Christensen, G.D. (1988). Characteristics of Adherence to Plastic Tissue Culture Plates of Coagulase-Negative Staphylococci Exposed to Subinhibitory Concentrations of Antimicrobial Agents. Journal of Infectious Diseases *157*, 71-77.

Shi, W., Jewett, A., and Hume, W.R. (1998). Rapid and Quantitative Detection of Streptococcus mutans with Species-Specific Monoclonal Antibodies. Hybridoma *17*, 365-371.

Spoering, A.L., and Lewis, K. (2001). Biofilms and Planktonic Cells of Pseudomonas aeruginosa Have Similar Resistance to Killing by Antimicrobials. Journal of Bacteriology *183*, 6746-6751.

Stewart, P.S., Mukherjee, P.K., and Ghannoum, M.A. (2004). Biofilm Antimicrobial Resistance. In Microbial Biofilms (American Society of Microbiology), pp. 250-268.

Stone, G., Wood, P., Dixon, L., Keyhan, M., and Matin, A. (2002). Tetracycline Rapidly Reaches All the Constituent Cells of Uropathogenic Escherichia coli Biofilms. Antimicrobial Agents and Chemotherapy *46*, 2458-2461.

Thomas, J.G., Lopez-Ribot, J.L., and Ramage, G. (2004). Biofilms and Implant Infections. In Microbial Biofilms (American Society of Microbiology), pp. 269-293.

Tichy, H.V., Wiesner, P., and Simon, R. (1996). PCR-Fingerprint-Verfahren zur Analyse von Mikroorganismen-Populationen. In Ökologie der Abwasserorganismen (Springer Berlin Heidelberg), pp. 111-122.

Timur, S., Haghighi, B., Tkac, J., Pazarlıoğlu, N., Telefoncu, A., and Gorton, L. (2007). Electrical wiring of Pseudomonas putida and Pseudomonas fluorescens with osmium redox polymers. Bioelectrochemistry *71*, 38-45.

Tomaras, A.P. (2003). Attachment to and biofilm formation on abiotic surfaces by Acinetobacter baumannii: involvement of a novel chaperone-usher pili assembly system. Microbiology *149*, 3473-3484.

van der Mei, H.C., and Busscher, H.J. (2001). Electrophoretic Mobility Distributions of Single-Strain Microbial Populations. Applied and Environmental Microbiology *67*, 491-494.

Vostiar, I., Ferapontova, E.E., and Gorton, L. (2004). Electrical “wiring” of viable Gluconobacter oxydans cells with a flexible osmium-redox polyelectrolyte. Electrochemistry Communications *6*, 621-626.

Walters, M.C., Roe, F., Bugnicourt, A., Franklin, M.J., and Stewart, P.S. (2003). Contributions of Antibiotic Penetration, Oxygen Limitation, and Low Metabolic Activity to Tolerance of Pseudomonas aeruginosa Biofilms to Ciprofloxacin and Tobramycin. Antimicrobial Agents and Chemotherapy *47*, 317-323.

website, U.K.M.F. http: depts.washington.edu/keck/intro.htm, Confocal introduction.

William Costerton, J. (2004). A Short History of the Development of the Biofilm Concept. In Microbial Biofilms (American Society of Microbiology), pp. 4-19.

Wolfaardt, G.M., Lawrence, J.R., Robarts, R.D., and Caldwell, D.E. (1994). The role of interactions, sessile growth, and nutrient amendments on the degradative efficiency of a microbial consortium. Canadian Journal of Microbiology *40*, 331-340.

Wright, C.J., Shah, M.K., Powell, L.C., and Armstrong, I. (2010). Application of AFM from microbial cell to biofilm. Scanning *32*, 134-149.

Zhang, T.C., Fu, Y.-C., and Bishop, P.L. (1995). Competition for substrate and space in biofilms. Water Environment Research *67*, 992-1003.

Zheng, Z., and Stewart, P.S. (2002). Penetration of Rifampin through Staphylococcus epidermidis Biofilms. Antimicrobial Agents and Chemotherapy *46*, 900-903.

Mechanisms of Biofilm Formation in Clinically used Biomaterials

4

John-Jairo Aguilera-Correa, Jaime Esteban* and David Romera-García

Department of Clinical Microbiology, IIS-Fundacion Jimenez Diaz, Madrid, Spain

*jestebanmoreno@gmail.com

DOI: https://doi.org/10.21775/9781912530328.04

Abstract

Medical-device infections are associated with high morbidity and exhibits robustness typically due to formation of biofilms on the device surface and surrounding tissues. Biofilms are complex communities composed of microorganisms embedded in secreted polysaccharide matrix which commonly display high resistance against the host immune system and antimicrobial therapy and can lead to a systemic infection or device failure leading to high costs in the healthcare. Frequently isolated agents in medical devices include bacteria such as *Staphylococcus aureus* and *Pseudomonas aeruginosa* and fungi represented mainly by *Candida*. In this chapter, we provide an overview of the current types of clinically used biomaterials, explore the major strategies and mechanisms developed by microorganisms to colonize devices, initial forces that are responsible for the interaction among the membranes of the microorganisms and the biomaterial surface and conditioning layer and regulatory systems that govern biofilm dispersal, expression of virulence factors and immune response evasion.

Introduction

Use of indwelling medical devices in health centers have greatly enhanced the healthy and quality of life of patients around the world. These devices include contact lenses, central venous catheters, needleless connectors, endotracheal tubes, intrauterine devices, mechanical heart valves, pacemakers, peritoneal dialysis catheters, prosthetic joints, tympanostomy tubes, urinary catheters, and voice prostheses, and many others (Table 1). Medical devices may be made of a wide range of biomaterials, from metallic alloys to ceramics and from polymers to hybrid composites.

These medical devices are made of at least one biomaterial. Biomaterial is synthetic or semi-synthetic material used to replace part of a living system or to function in intimate contact with living tissue (Geetha et al., 2009). The characteristics to be taken into consideration when it is meant as a biomaterial for implant include biocompatibility (non-toxic, tissues and blood compatible), biofunctionality (the biomaterial must perform the functions to which it has been designed for, at the required period of time and effectively), chemical and mechanical stability (dos Santos et al., 2017). Furthermore, the biomaterial should be sterilizable (dos Santos et al., 2017). Clinically used biomaterials are mostly of two types: metallic biomaterials and polymeric biomaterials. These materials differ widely in both chemical and mechanical features, but they share properties that make them ideal for use in specific clinical applications, such as an appropriate modulus of elasticity, resistance to wear and tear, and propensity for tissue integration (Rochford et al., 2012).

Metallic biomaterials are widely used in orthopedic implants, bone fixators, artificial joints, external osteosynthetic fixators and dental implants since they substitute for the functions of hard tissues in orthopedics and dentistry (Hanawa, 2012). Metallic biomaterials mainly include austenitic stainless steels, cobalt-chromium alloys and titanium alloys (Long and Rack, 1998). Other metal alloys of magnesium base, such as magnesium-zinc-calcium, magnesium-yttrium-zinc and magnesium-dysprosium are currently being studied as potential use materials in the manufacture of resorbable internal fixation implants (Malekani, 2012).

Polymeric biomaterials are widely used in all branches of Modern Medicine (Anesthesiology, Cardiovascular surgery, Dentistry, Otorhinolaryngology, Gastroenterology, Urology, Hematology, Neurology, Plastic surgery,

Table 1. Types of medical devices based on function (modified from (von Eiff et al., 2005) according to (Deva et al., 2013; Donlan, 2001)).

Intravascular
Peripheral catheters (venous, arterial)
Midline catheters
Central venous catheter
Non-tunneled catheters
Tunneled catheters
Pulmonary artery catheters
Totally implanted ports
Cardiovascular
Mechanical heart valves
Implantable defibrillators and related devices
Vascular grafts
Ventricular assist devices
Coronary stents
Implantable patient monitors
Neurosurgical
Ventricular shunts
Ommaya reservoirs
Intracranial pressure devices
Implantable neurological stimulators
Digestive
Peritoneal dialysis catheters
Biliary stents
Meshes
Orthopedic
Joint prostheses
Fracture fixation devices
Spinal implants
Fracture-fixation devices

Urological	
Inflatable penile implants	
Plastic surgery	
Breast implants	
Otolaryngologic	
Cochlear implants	
Middle ear implant	
Voice prostheses	
Tympanostomy tubes	
Ophthalmological	
Intra-ocular lenses	
Glaucoma tubes	
Dental	
Dental implant	
Denture	

Ophthalmology and Orthopedics) and play very different functions in each one of them (Teo et al., 2016). Most used polymeric biomaterials include polyethylene, polytetrafluoroethylene, polyamide, polypropylene, polyethylene-terephthalate, polydimethylsiloxane, polyhydroxyalkanoates, and polyetherether-ketone (He and Benson, 2014).

Infection can result from use of clinically used biomaterials (Donlan, 2001) and all patients receiving implanted medical device are at increased risk of developing an infection in the surgery site (Rochford et al., 2012). These medical-device infections affect the quality of healthcare in terms of increased morbidity, mortality, and additional cost for patient care (Iordanou et al., 2017). This kind of infection is associated with the microbial biofilm development on the medical-device surfaces. A biofilm is a bacterial aggregate that is reversibly or irreversibly associated with a surface or an interface and is embedded in a matrix of extracellular polymeric substances produced by themselves (Costerton et al., 1995), where numerous and complex socio-microbiological interactions occur (Diggle, 2010; Nadell et al., 2009; West et al., 2006). These composition of these communities are approximately 10% microorganisms and rest is biomatrix (Flemming et al., 2016). The development of the biofilm is one of the most widely distributed, successful (Flemming et al., 2016) and ancestral forms of growth on earth (Hall-Stoodley et al., 2004). Over 99% of the bacteria so far described are able to persist in biofilms which can provide an advantage living in this state (Vickery et al., 2013). From the clinical point of view, the importance of biofilms resides in its great antibiotic and immune system resistance (Monroe, 2007).

The focus of this chapter is on biofilm development formed on clinically used biomaterials. Initially, the most common bacteria that are responsible for the majority of these kind of infections, especially infections associated with medical devices are detailed according to their category including Gram nature and yeast. The final section of this chapter is dedicated to biofilm development on medical devices that manifest acute-infection signs after a long time of useful life.

Biofilm development

In-vitro studies have described biofilm formation as a complex process consisting in distinct stages i.e. adhesion, aggregation, maturation, detachment and dispersal. Adhesion is a two-step stage which involves the identification of surface by microorganisms followed by reversible and irreversible adhesion. The reversible adhesion is mediated by non-specific forces (Lifshitz-van der Waals, Lewis acid-base and electrostatic forces) (Bos et al., 1999) with bacteria behaving like colloidal microparticles (Arciola et al., 2018). Though, these forces could predict the reversible adhesion mediated to the extended Derjaguin-Landau-Verwey-Overbeek theory (Nguyen et al., 2011), they do not rightly predict the viable bacterial behavior, since this adhesion depends on the surface properties and bacterial features, such as species, strain, heterogeneity and cycle phase (Arciola et al., 2018). Irreversible adhesion occurs due to specific adhesions present on the pili, fimbriae, fimbriae-like appendages or cell surface. Maturation involves the aggregation and multiplication of bacteria on surface after adhesion to form three-dimensional microcolonies (Joo and Otto, 2012). The irreversible adhesion to biomaterials leads to change in gene expression, resulting in the synthesis and secretion of extracellular polymeric matrix (EPS) which acts as embedding substance and holds bacteria. The last stage of biofilm development includes the detachment of microorganisms from three-dimensional colonies, their translocation or dispersal and again adhesion to new location. These stages will be reviewed and described specifically for each bacterial species later.

The *in-vivo* appearance and structure of biofilms associated with biomaterials differ from those observed *in vitro*. Thus, these biofilms lack large three-dimensional structures, contain new components resulting from the mixture of the biofilm with biomolecules of the host, and present in a form of small colonies and heterogeneously distributed throughout the implant (Bjarnsholt et al., 2013). Furthermore, the EPS composition of *in-vivo* biofilms may be enriched by the host biomolecules and cells (Bjarnsholt et al., 2013) and the dispersion is dependent on different forces such as shear and pressure on medical devices that are implanted in the human body (Hall-Stoodley et al., 2004).

The biofilm development on clinically used biomaterials can be described a five-steps process which includes the formation of an initial conditioning layer, reversible and irreversible bacterial adhesion, a population growth or maturation, and an eventual detachment from a mature biofilm for a subsequent colonization (Garrett et al., 2008).

The conditioning layer of clinically used biomaterials

The conditioning layer is composed of organic and or inorganic compounds which settle on the biomaterial and modify their surface characteristics, such as surface charge, potential and tension (Jaffer et al., 2015). Thus, this layer act as an anchorage surface where cells and bacteria can adhere and proliferate.

There are different types of medical devices depending on its levels of invasiveness in the body, the anatomical site of insertion and the time of use, therefore these devices can be classified in transient, short-term, long-term or permanent devices (Arciola et al., 2018). Thus, depending on anatomical site of insertion, the conditioning layer will be composed by different biomolecules, e. g. the conditioning layer of an intravenous catheter or joint protheses are more similar than the layer of an endotracheal intubation tube or urinary catheter. In addition to implantation site, the exact composition of this conditioning layer depends on device surface chemistry (charge and hydrophobicity), topography, and exposure time (Rochford et al., 2012). The most studied formation of conditioning layer is on the invasive medical devices. In these medical devices, no matter where implants place, they will always be a foreign body (Arciola et al., 2018).

When blood covers a foreign body, one of the first events is the adsorption of blood proteins on the biomaterial surface. This protein adsorption can be detected as early as one minute on all materials (Anderson et al., 1990). Protein adsorption is followed by the Vroman effect, that involves the competition between protein adsorption and desorption (Vroman, 1988) e.g. high molecular weight kininogen and Hageman factor (factor XII) which are presented in scarce amounts in the blood and adsorb on the biomaterial surface in detriment of early adsorbed proteins (Anderson et al., 1990). Thus, the final conditioning layer on biomaterial used in indwelling device is composed of albumin, fibrinogen, complement (mainly C5), fibronectin, vitronectin, γ globulin, (which are the most abundant proteins), Hageman factor, high weight kininogen, factor VIII/vWF, hemoglobin (which are presented in moderated amounts), and control agents (which are in the lowest amounts) (Anderson et al., 1990; Anderson et al., 2008; Jaffer et al., 2015). Types, level, and surface conformations of adsorbed proteins are critical determinants for the peri-implant tissue reaction (Anderson et al., 2008) and are dependent on the characteristics of the biomaterial surface (Gristina, 1987). The final compositions and the role of the adsorbed proteins on the biomaterial surface will differ for each organism or type of tissue cells depending on the implantation site of medical device (Gristina, 1987).

After protein adsorption on biomaterial surface, blood cells such as erythrocytes, platelets and leukocytes attach to it. The most abundant leukocytes are neutrophils that prime the biomaterial surfaces through the release of chemical factors and neutrophil extracellular traps, which consists of web-like structures and composed of decondensed chromatin and antimicrobial factors (Brinkmann et al., 2004). The complement-leukocytes interaction results in clotting and inflammation (Anderson et al., 2008). This inflammation is firstly acute inflammation orchestrated by neutrophils that will degranulate giving rise to chronic inflammation which is characterized by monocytes/macrophages recruitment (Jaffer et al., 2015) and the tissue reparation forming a fibrous capsule over time. Neutrophils can influence the immune response either towards inflammation or resolution depending on the dominant neutrophil phenotype (Arciola et al., 2018). Thus, N1 phenotype is pro-inflammatory and increases expression of tumor necrosis factor, and the N2 phenotype is anti-inflammatory and produces more vascular endothelial growth factor (VEGF) and matrix metalloproteinase-9 which are important for angiogenesis (Selders et al., 2017).

The reasoning behind fibrous capsule formation is that if the implanted medical device cannot be ingested by macrophages, then the next best protection for the host seems to be the isolation of the foreign body using this capsule as a physical barrier that decreases host-biomaterial interaction (Nuss and von Rechenberg, 2008).

Infection on clinically used biomaterials: microbial adhesion and maturation

Biofilms on medical devices may be composed by different types of microorganisms. These microorganisms may originate from the skin of patients or health-workers, tap water to which entry ports are exposed, or other environmental sources (Donlan, 2001). The most frequently associated microorganisms to medical device infections are *Staphylococcus*, as the most frequent Gram-positive bacteria, followed by Gram-negative bacilli and finally and very uncommonly *Candida* yeasts. Most of them opportunistically infect hosts during invasive medical intervention, chemotherapy or via a pre-existing underlying disease (Hall-Stoodley et al., 2004) (Table 2).

Table 2. Major bacterial species associated to medical-device infections (modified from (Arciola et al., 2018) according to ¥: (Deorukhkar and Saini, 2016; Kojic and Darouiche, 2004)).§: These data can change in medical devices such as ventilator associated devices (25−39%) and urinary catheters (21%) (Kojic and Darouiche, 2004).

Microorganism		**Prevalence (%)**
	Species	
Bacteria		
	Staphylococcus aureus	13.0−31.0
	Coagulase-negative staphylococci	20.2−39.3
	Streptococci	10.3−14.5
	Escherichia coli	1.6−2.4
	Pseudomonas aeruginosa	6.1−6.7
Yeast		
	Candida spp.¥	0.5-9.2§

Gram-positive bacteria

Gram-positive bacteria are responsible of 40-80% of all medical device associated infections (Arciola et al., 2018; Benito et al., 2016) and are mostly represented by *Staphylococcus aureus,* Coagulase-negative staphylococci (CoNS), *Streptococcus spp.* (including oral streptococci such as *S. mutans* or *S. agalactiae*) and *Enterococcus spp.* (with *E. faecalis* and *E. faecium* as the most frequently isolated species). CoNS are isolated as the first cause of infection in peripheral and central venous catheter, central line and in periprosthetic joint infections (Huebner and Goldmann, 1999; Rogers et al., 2009), whereas *S. aureus* is isolated as the first cause in infections involving internal or external fixation (Arciola et al., 2005). Up to 80% of CoNS identified in these cases were *S. epidermidis* (Donlan, 2001). Additionally, streptococcus and enterococcus are responsible of 10.3-14.5% of implant infections (Aggarwal et al., 2014).

Streptococcal biofilms have been mostly studied in the context of oral infections and its formation over implants (Karchmer and Longworth, 2002) and infections caused by enterococcus are mainly catheter-associated infections, and are responsible for 6% of catheter-associated urinary tract infections (Melzer and Welch, 2013; Sandoe et al., 2003) and 10% of catheter related bloodstream infections in patients receiving home parenteral nutrition (Nielsen et al., 2012).

Infections caused by these microorganisms originate with the introduction of the pathogen from normal flora of the patient or health personnel during a medical procedure, trauma, or opportunistically in immunocompromised patients (Alberici et al., 2015; Lee et al., 1999; O'Grady N et al., 2002; Otto, 2009). Medical device infections caused by *S. aureus* are often early acute infections with high inflammation and important symptoms, whereas most CoNS are considered low virulent pathogens that cause late infections. The isolated strains tend to display an elevated resistance to antibiotics such as methicillin or gentamicin (Campoccia et al., 2006; Diekema et al., 2001). It has been described that in some areas up to 40-60% of hospital-acquired *S. aureus* strains are methicillin-resistant strains and there is an increasing concern of antibiotic resistance in CoNS such as *S. epidermidis.* (Diekema et al., 2001). Initial attachment to medical device surfaces as well as the conditioning layer is mediated by expression of surface factors that bind host proteins like fibronectin or fibrinogen, such as autolysins *AtlA* (Foster, 1995) in *S. aureus*/AtlE in *S. epidermidis* (Heilmann et al., 1997), Ssp-5 (serin-rich family) (Tettelin et al., 2002), of *S. agalactiae* or protein antigen I/II (Beg et al., 2002) in *S. mutans;* and adhesins (Lowe et al., 1995) and the cell-wall-associated rhamnopolysaccharide *Epa* (enterococcus polysaccharide antigen) (Guzman et al., 1991) in *E. faecalis.* Additionally, filamentous structures such as fimbria-like components SSP-1 and SSP-2 in *S. epidermidis* (Huebner and Goldmann, 1999; Veenstra et al., 1996), fibrils and adhesive pili in streptococci (Handley et al., 1987; Lauer et al., 2005) and enterococci (Nallapareddy et al., 2006) have been described to play a role in initial attachment to medical devices surfaces. Extracellular DNA has showed to play also a role in the attachment of several gram-positive bacteria such as *S. mutans* by changing the hydrophobicity of their membrane and allowing to interact with hydrophobic surfaces (Das et al., 2011). Interestingly, has been reported that teichoic acids are essential for biofilm formation and primary attachment to abiotic surfaces in *S. aureus* by maintaining surface net charge that will contribute to the hydrophobic interaction (Gross et al., 2001).

Next, irreversible attachment to host proteins and plasma clot is mediated by the expression of a vast repertoire of MSCRAMMs (microbial surface components recognizing adhesive matrix proteins) (Nobbs et al., 2009; Patti and Hook, 1994; Sillanpaa et al., 2008) and SERAMs (Secretable expanded repertoire adhesive molecules) (Foster et al., 2014). Relevant examples are *Bap* (biofilm-associated protein) (Latasa et al., 2006) and surface protein G *SasG* (Kuroda et al., 2008) of *S. aureus;* fibrinogen-binding protein *SdrG/Fbe* of *S. epidermidis* (Speziale et al., 2014), glucan binding proteins (glucan polysaccharides that bind to the salivary pellicle) of *S. mutans* (Banas and Vickerman, 2003) or the collagen/ laminin binding protein Ace of *E. faecalis* (Sillanpaa et al., 2004).

Bacterial accumulation and aggregation are mediated by specific proteins and accessory factors such as adhesins or fimbriae/pili. Specific proteins Accumulation-associated protein *Aap* (McCann et al., 2008) of *S. epidermidis;* aggregated signal peptide (AgrD*)* (Schwartz et al., 2014) and phenol-soluble modulins (Schwartz et al., 2012) in *S. aureus,* whose forms amyloid fibrils on the surface of the bacteria which contribute to the intercellular aggregation; proteins of the family AgI/II (Kelly et al., 1989) and glucan-binding proteins in streptococci; the aggregation substance (*Agg*) (Sussmuth et al., 2000), *Esp* (Toledo-Arana et al., 2001) and gelatinase *gelE* (Engelbert et al., 2004) of *E. faecalis.*

Here as main extracellular matrix component, staphylococci produce and incorporate PNAG (poly-N-acetylglucosamine) and PNSG (poly-N-succinylglucosamine), also called polymeric intercellular adhesin, (PIA) (Mack et al., 1996a), which facilitates interaction between glycosaminoglycans and cell-wall components, polysaccharides or lectins, all of which leading to the bacterial accumulation and increasing complexity of the biofilm structure (Lin et al., 2015; Mack et al., 1996b). Moreover, PIA production has been linked with enhanced antibiotic resistance (O'Gara, 2007). Additionally, isolates have been described (especially MRSA) that develop biofilms in a PIA-independent way and instead produce highly FnBP-dependent biofilms in medical device infections induced in *in vivo* models (McCourt et al., 2014; O'Neill et al., 2008). Exceptionally, MRSA isolates can incorporate host fibrin as main component into the ECM in a PIA-independent biofilm development (Dakheel et al., 2016; Geoghegan et al., 2013; O'Neill et al., 2008). Streptococci produce polypeptide capsules that contribute to attachment to host tissues (Jenkinson and Demuth, 1997), and an extracellular matrix composed mostly of glucans synthesized by streptococcal glucosyltransferases (Gtfs) (Loesche, 1996). For example, *S. mutans* glucan polysaccharides contribute to establish interactions with other species of the oral bacteria (Banas and Vickerman, 2003). Most beta-hemolytic streptococci produce a hyaluronic acid made capsule and also play a role as virulence factor (Brown et al., 1983). In enterococci, some studies have suggested a structural role for anionic polysaccharides (Erlandsen et al., 2004) and *Epa* (Teng et al., 2009). The production of PIA is regulated by the *icaABCD* operon (Mack et al., 2000), which is regulated by stress conditions such as oxygen limitation, presence of antibiotics or temperature. In enterococci, quorum system *fsr* was identified and was described as a homologue to the *agr* system in staphylococci which regulates biofilm formation, maturation and production of virulence factors (Qin et al., 2000). Most relevant virulence factors produced by Gram-positive bacteria are summarized in Table 3.

Extracellular DNA (eDNA) is another major component in the extra cellular matrix of biofilms composed of Gram-positive bacteria. This ECM has several functions such as attachment to biotic surfaces or intercellular attachment (Das et al., 2013; Whitchurch et al., 2002). Programmed cell-lysis of small groups of bacteria is the most common mechanism for eDNA production (Jung et al., 2017; Turnbull et al., 2016). Recently, De Francesco *et al.* described in *S. aureus* a mechanism where membrane destabilization is essential (DeFrancesco et al., 2017). Importantly, most of the above-mentioned factors that contribute in early attachment and bacterial aggregation have additional roles in other processes during biofilm formation. For example, staphylococcal autolysins and hydrolases participate in promoting cell lysis, eDNA release and cell accumulation (Bose et al., 2012). Extracellular DNA have additional

Table 3. Virulence factors of Gram-positive bacteria.

Species	Protein or virulence factor	Effects to host during infections	References
CoNS (*S. epidermidis*)	SepA protease	AMP degradation	(Lai et al., 2007)
	APS system	AMP sensing system	(Li et al., 2007)
	Poly-γ-glutamic acid (PGA)	Protection against AMP and phagocytosis	(Kocianova et al., 2005)
	PSMs (α, β, δ, ε)	Pro-inflammatory cytolisins	(Vuong et al., 2004a)
	Glutamylendopeptidase and serine protease (GluSE, SsA and Esp)	Fibrinogen and complement factor C5 degradation	(Dubin et al., 2001)
	PNAG/PIA	Prevention from neutrophil attacking and opsonization	(Vuong et al., 2004c)
S. aureus	PSMs	Pro-inflammatory cytolisins, cell apoptosis (in MRSA)	(Joo et al., 2010)
	Enterotoxins*	Massive activation of T-cells and antibody presenting cells	(Geiger et al., 2012)
	Toxic shock syndrome toxin-1*	Massive activation of T-cells and antibody presenting cells	(Lin and Peterson, 2010)
	Leucocidins	Leucocytes lysis	(Szmigielski et al., 1999)
	Staphopain B	Tissue damage	(Dinges et al., 2000)
	Aureolysin	AMP resistance and tissue damage	(Laarman et al., 2011)
	Staphylokinase	Activate plasminogen; inactivate antimicrobial peptides	(Bokarewa et al., 2006)
	Protein A	Proinflammatory properties, prevents opsonization	(Graille et al., 2000)
Streptococci	Bacteriocins	Immunogenic (*S. mutans*)	(Kuramitsu, 1993)
	Capsule	Protection against immune system attacks. Endothelian cell invasion, pro-inflammatory	(Yother, 2011)
	Superantigens	Massive activation of T-cells and antibody presenting cells	(Fraser and Proft, 2008)
	Pili/fimbriae	Phagocyte resistance (*S. aalactiae*)	(Maisey et al., 2008)
Enterococci	Gelatinase	Hydrolysis of collagen and casein	(Singh et al., 2005)
	Cytolysins/hemolysins	Hemolysis/cytolysis	(Chow et al., 1993)

functions such as facilitating gene transfer (Roberts and Kreth, 2014), modulation of the immune response (Thurlow et al., 2011), and nutrients supply (Vorkapic et al., 2016), and PSMs contribute to the structure of the water channel in *S. epidermidis* (Le et al., 2014). AS of *E. faecalis* also mediates attachment to extracellular matrix proteins of the host (Rozdzinski et al., 2001), and increase survival inside polymorphonuclear leukocytes and macrophages (Rakita et al., 1999).

In *S. mutans* and *S. oralis,* anchorless-adhesins (similar to SERAMs) act as virulence factors and bind to plasminogen/plasmin (Kinnby et al., 2008). Moreover, in periprosthetic joint infections, expression of MSCRAMMs and PSMs was enhanced in presence of synovial fluid and human plasma (Cardile et al., 2014; Dastgheyb et al., 2015); hence, biofilms here present a more densely packed structure composed of peptidoglycan and matrix host proteins, collagen and fibronectin, which makes the treatment of these infection challenging. It has been suggested a role in later infection processes for SERAMs such as osteomyelitis (Kalinka et al., 2014). The whole process of Gram-positive biofilm formation and maturation is regulated by quorum system and additional regulator systems (Bassler, 1999; Flemming et al., 2016). After initial attachment, biofilm formation in streptococci is characterized by the induction of competence by the production of competence stimulating peptide *CSP* (Cvitkovitch, 2001). This is regulated by quorum sensing *comAB*, *comCDE* (Lee and Morrison, 1999) and *luxS* systems (also known as S-ribosyl-homocysteine lyase) (Merritt et al., 2003; Stroeher et al., 2003), that control the production of virulence factors together through the *Blp* system (Lau et al., 2001). Additionally, in *S. pneumoniae* nearly 30 regulatory systems controls biofilm formation and maturation and the regulates production of virulence factors (Shepherd et al., 2012).

In staphylococci, quorum systems *agr* and *luxS* regulate gene expression in function of cell density and environmental conditions. In *agr* system, protein AgrB is responsible for the production of autoinducer peptides (AIP), which are thiolactone-containing pheromone peptides whose primary sequence varies among staphylococcal strains, that initiate the signal cascade (Novick, 2003;

Otto, 2001). On the other hand, LuxS synthesize AI-2 (4,5-dihydroxy-2,3-pentanedione) (Winzer et al., 2002; Xu et al., 2006), which is considered a universal molecule that participates in inter-species communication (Xavier and Bassler, 2003).

During the stationary phase of biofilm formation and maturation, the *agr* system upregulates the production of molecules involved in detachment and interaction/ interference with immune system, whereas the expression of molecules related to attachment and interaction with the host tissues are down-regulated (Vuong et al., 2003; Vuong et al., 2004b). In fact, as toxins production is under very strict control of the *agr* system, it has been proposed to be considered as the regulator of toxins rather than a population controller in staphylococci (Novick, 2003).

Moreover, in septic failure, a small proportion of *S. aureus* and some species of CoNS have the capacity of hiding inside non-professional phagocytic cells (Campoccia et al., 2016; Maali et al., 2016) to avoid the action of antibiotics and also can persist inside osteoblasts and later induce apoptosis and bone destruction (Alexander et al., 2003). In addition, *S. aureus* can migrate though the canaliculi of bone until the osteocyte lacunae, where it can form biofilms, which would serve to evade the immune responses and antibiotics. Additionally, internalized bacteria in chronic medical device infections can develop a small-colony variant (SCV) phenotype, characterized by being hyper-piliated hyper-adherent, mostly auxotrophic for either menadione or hemin and exhibit high biofilm forming capacities with high auto-aggregative behavior (Proctor et al., 2006). In addition, SCV display augmented resistance to several classes of antibiotics and display low rates of growing and cytotoxic factor production (Bui et al., 2017), which is problematic because SCV are often missed during current culture methods for identification and can resist antibiotic therapy, especially in CVC infections (Neut et al., 2007).

Gram-negative bacteria

Whilst Gram-positive medical-device infections originates from patient's skin flora, Gram-negative ones are much less common and are related to endogenous and exogenous microbiota from several potential sources, including the patient and health-care personnel, or contaminated infuses, tap water to which entry posts are exposed originated from contaminated intravenous fluids (Donlan, 2001). Aerobic Gram-negative bacilli are more common in early-onset medical-device infections (Tande and Patel, 2014). These bacteria play a prominent role in the polymicrobial infections and may cause hematogenous monomicrobial infections (Benito et al., 2016). Between this kind of bacteria, there are two Gram-negative types: fermenting Gram-negative bacteria (mainly *Enterobacteriaceae*, represented by *Escherichia coli*) and non-fermenting Gram-negative bacteria (mainly *Pseudomonas aeruginosa*). The medical-device infections provoked by Gram-negative are increasing over the last years in specific device such as joint prostheses (Benito et al., 2016) or venous catheters. Hence, *Enterobacteriaceae* family and *P. aeruginosa* may represent 29.2 % of the prosthetic joint infections (Benito et al., 2016), 31.3 % of invasive device (ventilation device, central bloodstream and urinary catheters) infections taking place in intensive care units (Bianco et al., 2018), and up to 80% of central catheter infection in oncological patients (Patel, 2016). Whilst the primary pathogens involved in the urinary catheter infection are mainly *Enterobacteriaceae* like *E. coli*, *Proteus mirabilis*, and *Klebsiella pneumoniae* (Nicolle, 2014), *P. aeruginosa* is associated up to 25% of ventilator-associated pneumonia (Iordanou et al., 2017).

Gram-negative irreversible attachment mainly depends on the hydrophobicity of biomaterial surface and the appendages such as flagella and pili (Garrett et al., 2008). Flagella are involved in the irreversible attachment on abiotic surface in *E. coli* (Friedlander et al., 2015) and on super hydrophobic surface in *P. aeruginosa* (Bruzaud, 2015). Furthermore, type IV pili are involved in the adherence on abiotic surface in *P. aeruginosa* (Giltner et al., 2006) and in *E. coli* (Laverty et al., 2014). In *P. aeruginosa*, the O-polysaccharide chain of the lipopolysaccharide modulates the hydrophobicity of its outer membrane, since the bacterial outer surface is more hydrophobic if lipopolysaccharide A predominates and is more hydrophilic if lipopolysaccharide B predominates (Laverty et al., 2014). Moreover, eDNA and the Psl exopolysaccharide are associated with *P. aeruginosa* adhesion (Joo and Otto, 2012), while curli and the polysaccharide cellulose promotes *E. coli* adhesion (Hufnagel et al., 2015). Moreover, the presence of divalent cations, such as Zn^{2+}, Ca^{2+} and Mg^{2+}, and changes in urinary pH and ionic strengths all resulted in an increase in bacterial attachment in the surface of medical devices infections (Nicolle, 2014).

After irreversible attachment, accumulation and ultimately maturation of Gram-negative biofilm correspond to the increased productions of EPS and the downregulation of the genes required for motility apparatus (Laverty et al., 2014). The quorum sensing plays a crucial play in the maturation of the Gram-negative biofilm development (Laverty et al., 2014). This quorum sensing mechanism depends on signaling molecules derivates from *N*-acyl-homoserine lactone. Four different molecules associated with quorum sensing have been described for *P. aeruginosa*: *N*-(3-oxododecanoyl)-L-homoserine lactone (Passador et al., 1993), *N*-butyl-homoserine lactone (Ochsner and Reiser, 1995), 2-heptyl-3-hydroxyl-4-quinolone (Diggle et al., 2006) and LasR-*N*-(3-oxododecanoyl)-L-homoserine lactone (Latifi et al., 1996). To date, there are three identified quorum sensing systems (Laverty et al., 2014). These are the *las*-based system controlled via *N*-(3-oxododecanoyl)-l-homoserine lactone production by *lasRI* gene loci (Passador et al., 1993); the *rhl* system regulated by *N*-butyryl-homoserine lactone produced by the *rhlRI* operon (Ochsner and Reiser, 1995) and the *pqs* system that controls the las and *rhl* quorum sensing through 2-heptyl-3-hydroxy-4-quinolone production (Diggle et al., 2006). The *rhl* system is controlled by the las system as LasR-*N*-(3-oxododecanoyl)-l-homoserine lactone activates the transcription of *rhlR* and *rhlI* which indicates that these two systems are interlinked (Latifi et al., 1996).

In a similar way, four mechanisms of quorum sensing have been described for *E. coli* controlled by four different molecules: N-acyl-1-homoserine lactone (Ahmer, 2004; Van Houdt et al., 2006), autoinducer-2 (De Keersmaecker et al., 2006), autoinducer-3 (Sperandio et al., 2003) and indole (Walters and Sperandio, 2006). *E. coli* quorum sensing is based on four different systems (Laverty et al., 2014): the LuxR homologue *sdiA* system (Ahmer, 2004; Van Houdt et al., 2006); the *luxS*/autoinducer-II system (De Keersmaecker et al.,

2006); an autoinducer-III/epinephrine/norepinephrine system (Sperandio et al., 2003), and an indole mediated system (Walters and Sperandio, 2006).

Quorum sensing regulates the expression of a huge plethora of genes related to biofilm development and maturation and the virulence factors of these species. Thus, this intercellular-communication mechanism upregulates exopolymeric matrix synthesis of Gram-negative biofilm. This biofilm matrix is composed of polysaccharides, proteins and extracellular DNA (Flemming and Wingender, 2010). One main difference between *E. coli* and *P. aeruginosa* biofilm matrix is the polysaccharide chemical composition.

Table 4. Virulence factors controlled by quorum sensing in Gram-negative bacteria. §: modified from (Lee and Zhang, 2015).

Species	Protein or virulence factor	Effects to host during infections	References
E. coli	α-haemolysin	Formation of pores in the cell membrane	(Goni and Ostolaza, 1998)
	Cytotoxic necrotising factor 1	Polymorphonuclear phagocytosis evasión and induction of apoptotic death of bladder epithelial cells	(Emody et al., 2003)
	Secreted autotransporter toxin (SAT)	Formation of lesions in the tight junctions and autophagy in epithelial cells	(Guignot et al., 2007; Lievin-Le Moal et al., 2011)
	Production of cytolethal distending toxin (CDT)	Block the cell cycle in G_2 phase.	(Toth et al., 2003)
	Cytolysin A	Formation of pores in the cell membrane	(Ludwig et al., 2004)
P. aeruginosa §	Elastase	Degradation of elastin, collagen, and other matrix proteins	(Wolz et al., 1994; Yanagihara et al., 2003)
	Protease	Disruption of epithelial barrier	(Kessler et al., 1993; Park et al., 2000)

Exotoxin A	Cell death	(Daddaoua et al., 2012)
Alkaline protease	Degradation of host complement system and cytokines	(Laarman et al., 2012)
Rhamnosyl-transferase (rhamnolipids)	Necrosis of host macrophage and polymorphonuclear lymphocytes	(Alhede et al., 2014; Jensen et al., 2006; Lequette and Greenberg, 2005)
Hydrogen cyanide	Cellular respiration arrest	(Adam et al., 1997; Ryall et al., 2008)
Pyocyanin	Oxidative effects dampen host cellular respiration and causes oxidative stress.	(Denning et al., 1998; Jackowski et al., 1991; Lau et al., 2004)

P. aeruginosa biofilm polysaccharide can be composed by Psl (Ma et al., 2007), Pel (Ryder et al., 2007) and alginate (Boyd and Chakrabarty, 1995) but they are not essential for biofilm development (Laverty et al., 2014). Psl polysaccharide is a galactose-rich and mannose-rich exopolysaccharide (Joo and Otto, 2012; Ma et al., 2007) and can contain glucose, rhamnose and a limited amount of xylose (Laverty et al., 2014). Pel polysaccharide is a positively charged exopolysaccharide composed of partially acetylated 1→4 glycosidic linkages of *N*-acetylgalactosamine and *N*-acetylglucosamine which plays a crucial role in the interaction and cross-linkage with other key *P. aeruginosa* biofilm matrix components such as eDNA (Jennings et al., 2015). Alginate is composed by a random polymerisation of mannuronate (M) and guluronate (G) units where the most common triad is MM and where GM and MG pairs may be a necessary structural element for bacterial alginate to form a stable gel in presence of ambient Ca^{2+} cations (Schurks et al., 2002). *P. aeruginosa* exopolysaccharide synthesis begin just after attachment (Chrzanowska, 2014; Fischer et al., 2004; Kanmani, 2013). The *E. coli* biofilm matrix is mainly composed of colanic acid.

Colanic acid consists of hexasaccharide subunits mainly of fucose and glucuronic acid (Obadia et al., 2007). It is to be noted that *E. coli* small colony variant strains upregulate the colanic acid synthesis genes compared to wild type strains (Santos and Hirshfield, 2016). Unlike *P. aeruginosa*, *E. coli* exopolysaccharide synthesis may be present prior to bacterial attachment on biomaterial surface (Huang et al., 2015; Hufnagel et al., 2015).

The virulence factors controlled by quorum sensing are numerous and varied in these two Gram-negative bacteria and they are summarized in the Table 4.

The eDNA presented in the Gram-negative bacterial biofilms originates from bacterial lysis and has specific functions. eDNA compensates structurally the lack of alginate in biofilms from non-mucoid *P. aeruginosa* strains (Ghafoor et al., 2011), play a very similar role as structural *E. coli* biofilm component (Hung et al., 2013) and chelates cationic antimicrobials (Laverty et al., 2014).

Three-dimensional structure and microcolony anatomy depend on at least two mechanisms: the above-mentioned quorum sensing mechanism that modulates the activity of the extracellular enzyme and the external forces depending on the biofilm localizations. Up to 19 different extracellular enzymes have been described with a protective role in *P. aeruginosa* (Tseng et al., 2018) and other ones which modify chemically the outer membrane lipopolysaccharides (Nuri et al., 2015) and remodel the biofilm matrix in both *E. coli* and *P. aeruginosa* (Laverty et al., 2014).

Yeast

Medical device infections caused by fungi represent 65-90% of denture and voice prostheses infections (Gendreau and Loewy, 2011; Talpaert et al., 2015), 8-16% of central venous catheter-associated infections (Pannanusorn et al., 2013) and 8% in catheter-associated urinary tract infections or 1-2% in prosthetic joint infections. The most common identified pathogen is *C. albicans* (Kojic and Darouiche, 2004), although other species has been identified such as *C. parapsilosis* and *C glabrata* (Silva et al., 2012) or the recently discovered *C. auris*, which has been identified in catheter associated infections in immunocompromised patients in intensive care units (Lee et al., 2011). The mortality rate associated with invasive *Candida* infections derived from device infections is reported as high as 45% (Wenzel, 1995).

It is important to note that in this section the process of biofilm formation will be described for *C. albicans*, as it has been characterized more exhaustively than other species such as *C. parapsilosis* or *C. glabrata* due to its clinical relevance. However, most of the factors mentioned for *C. albicans* have its orthologs present in the other species and display similar functions.

Initial attachment to biomaterial surface is mediated by specific individual cell wall adhesins such as *Eap1* (enhanced attachment to polystyrene) (Li and Palecek, 2008) and agglutinin-like sequence proteins *Als1* and *Als5.* These proteins contribute to the hydrophobic interaction between yeast and biomaterial surface and mediate binding to host proteins that cover the biomaterial surface covered by proteinaceous conditioning layer such as fibrinogen and fibronectin (Douglas, 2002; Henriques et al., 2006). The initial focal attachment of yeast cells to a biomaterial surface is closely followed by hyphae formation, proliferation and biofilm development (Ramage et al., 2006). Many of the proteins required for biofilm initiation are also required for hyphae formation. *Eap1* and other proteins such as *Als3*, *Hwp1, Hwp2* and *Rbt1* contribute to this process (Ene and Bennett, 2009), and once the hyphae are formed, *Hwp1* and *Als3* mediate cell-cell aggregation and contribute to cell-substrate binding (Nobile et al., 2006; Nobile et al., 2008).

After attachment and hyphae formation, biofilm matrix production is addressed. Mature biofilms formed by *C. albicans* are three-dimensional bilayer structures mainly composed of yeast-layer and hyphal/pseudohyphal layer forming microcolonies separated by water channels and embedded in extracellular matrix, which is formed by carbohydrates with β-1,3-glucan as major carbohydrate component, and other glycans such as β-1,6-glucan, proteins, phosphorus, hexosamines, phosphorus, uronic acid, and eDNA (Pierce et al., 2017). Here, composition varies among other *Candida* species. For example, biofilms formed by *C. albicans* consist of a dense network of yeasts, hyphae, and pseudohyphae with enhanced presence of extracellular matrix, whereas *C. parapsilosis, C. orthopsilosis* and *C. metasilopsis* are composed of regular groups of cells with minimal presence of extracellular matrix (Kuhn et al., 2002; Lattif et al., 2010).

During the different phases of biofilm development, genes encoding ergosterol biosynthesis, efflux pumps, major facilitator proteins, chitinases and heat shock/ stress response proteins are upregulated (Araujo et al., 2017; Murillo et al., 2005; Ramage et al., 2002a). Also, β-1,3 glucan matrix of Candida biofilms binds to and sequesters azole drugs (Mukherjee et al., 2003), and participate in the resistance against polyenes and echinocandins (Nett et al., 2010).

Similar to other bacterial systems, the process of biofilm formation in *Candida* is controlled by quorum system. Two main quorum systems were identified: E-E-farnesol (Hornby et al., 2001) and Tyrosol (Chen et al., 2004). Alem et al. demonstrated that during the first stages of biofilm formation, tyrosol stimulates initial adhesion and filamentation (Alem et al., 2006), whereas farnesol has the opposite effect and is involved in production of virulence factors and dispersal (Navarathna et al., 2007; Ramage et al., 2002b). Signal transduction of quorum sensing systems in *Candida albicans* has been suggested to be carried on by multiple factors (Kruppa, 2009) that is yet to be determined. This signal cascade will presumably culminate with the production of a number of well-known transcription factors involved in early adhesion to substrates and/or hyphal differentiation. An important example of these factors is biofilm and cell wall regulator *Bcr1* (Nobile and Mitchell, 2005), which is essential for the expression of proteins such as TEC1, involved in primary attachment, or EFG1, which contributes to the expression of cell surface proteins and hyphal formation (Srikantha et al., 2013; Yano et al., 2016).

Main virulence factors produced by *Candida* biofilms are summarized in Table 5. Moreover, phenotypic switching between yeast and hyphal forms, β-glucans from extracellular matrix and surface wall superoxide dismutase protect the yeast from ROS produced by neutrophils, and despite the fact that neutrophils and macrophages are able to phagocytose fungal cells, *C. albicans* is able to adapt to the phagocytic environment (production of ROS and AMPs, nutrient starvation) by switching their metabolism from glycolysis to glyoxylate cycle and glyconeogenesis (Frohner et al., 2009).

Table 5. Virulence factors of *Candida* species.

Species	Protein or virulence factor	Effects to host during infections	References
C. albicans	Hydrolases (phospholipases, proteases, lypases)	Hydrolysis of host matrix proteins	(Monod and Borg-von, 2002; Schaller et al., 2005)
	Hwp1 invasin and Als3 adhesin	Penetration and escape in neutrophils and PMNs by mechanical piercing	(Lorenz et al., 2004)

Bacterial detachment and dispersal

Detachment is the most general term to indicate the reduction of the total mass of biofilms and allows colonization of new niches (Hall-Stoodley et al., 2004) and trigger due to lack of nutrients, unfavorable conditions and external mechanical forces. The detachment can be classified into three different types (Kim and Lee, 2016): sloughing, erosion, and seeding dispersal. Sloughing is the process in which biofilm is stripped off as a lump from the biomaterial surface and causes a rapid loss of biofilm mass. In contrast, erosion is a process in which a single bacterium or small piece of biofilm is detached from outside. While these types of detachment are generally considered shear-dependent and passive processes, seeding dispersal is an active process and refers to the rapid release of single cells or small amounts of bacteria from the central region of biofilm, leaving hollow cavities. These types of detachment occur independently of the bacterial species involved in the infection and they can caused complications such as metastatic seedic (e.g. osteomielitis or abcesses), septic thrombophlebitis and thromboembolism (von Eiff et al., 2005).

Dispersal is generally considered a concomitant event with detachment, some exceptional dispersal events have been suggested such as rolling, rippling and swarming motion associated with type IV pili, in which dispersal does not seem absolutely associated with detachment and depend on the specific bacterial species involved in the infection (Kim and Lee, 2016) and their motility.

Biofilm dispersal is important for disseminating all bacteria; however, for the Gram-negative bacteria to leave the biofilm in which they are both protected and entrapped, it may be necessary to sacrifice part of the biofilm and have some cells undergo autolysis [203]. Therefore, programmed bacterial death may make sense for the biofilm and the primitive tissue that this collection of cells represents but not for planktonic cells.

In staphylococci, *agr* and *luxS* quorum systems regulate biofilm detachment and seeding dispersal through several mechanisms. For example, AI2 from *luxS* causes the reduction of cell-to-cell adhesion through the down-regulation of PIA expression (Xu et al., 2006), and *agr* is implicated in the production of proteolytic enzymes (Dubin et al., 2001) such as aureolysin (Boles and Horswill, 2008) and thermonuclease in *S. aureus* (Mann et al., 2009), and PSMs that act like detergents and degrade the biofilm exopolymers by disrupting covalent bonds (Le et al., 2014; Otto, 2013).

Dispersal of Streptococcal biofilms is poorly understood but some factors have been identified, such as the surface protein releasing enzyme (SPRE) which actively releases surface proteins implicated in attachment to surfaces in *S. mutans, S. pneumoniae* and *S. agalactiae* (Vats and Lee, 2000); and DeoC nuclease in *S. mutans* (Liu et al., 2017). Moreover, seeding dispersal has been documented in *S. mutans* controlled by *luxS* system (Kaplan and Fine, 2002).

Some *E. faecalis* probiotic strains has been observed seeding dispersal dependent on AI-2 synthesis by *luxS* quorum system, which induces prophage-mediated cell-lysis (Rossmann et al., 2015). Moreover, orthologues for these phage proteins have been identified in staphylococci and streptococci (Bayles, 2007; Rice et al., 2007), although the underlying mechanism and its importance in biofilm dispersal for these species has yet to be elucidated.

In *E. coli*, lysis via Hha may aid biofilm dispersal by forming holes in the biofilm matrix (Wood, 2009). The mechanism used by Hha is the control of translation via the availability of rare codon tRNAs which activates prophage lytic genes (Garcia-Contreras et al., 2008). One of the first sRNA systems discovered, sRNA binds a regulator protein to control the activity of the carbon storage regulation system that consists of CsrA. CsrA represses gluconeogenesis, glycogen metabolism, peptide transport and production of the adhesion poly-β-1,6-*N*-acetyl-D-glucosamine while it activates glycolysis, acetate metabolism, and flagella biosynthesis (Baker et al., 2007; Wood, 2009). CsrA also represses biofilm formation and increases biofilm dispersal (Jackson et al., 2002). Furthermore, exogenous glucose blocked *E. coli* biofilm dispersal induced by CsrA which further supports the hypothesis that nutrient cues can induce biofilm dispersal (Jackson et al., 2002). On the other hand, *E. coli* detachment also seems to be strongly influenced by sloughing and erosion according to recent reports (Oder et al., 2018).

In *P. aeruginosa*, quorum sensing controls EPS degrading enzymes and surfactants synthesis. Exopolysaccharide degrading enzymes including DNases, polysaccharide-degrading enzymes (e.g. alginate lyase), protease can cause the spatial differentiation and dispersion by destroying and modifying the biofilm matrix (Kim and Lee, 2016). The dispersal seems to be controlled by *las* and *rhl* quorum sensing systems (Purevdorj-Gage et al., 2005). Surfactants are amphipathic glycolipids such as rhamnolipids which are synthesized in the center of the mature biofilm structures and favor the chemical biofilm matrix disaggregation (Davey et al., 2003) and seeding dispersal. Rhamnolipids include three types of active molecules, di-rhamnolipids, mono-rhamnolipids, and 3-(3 hydroxyalkanoyloxy) alkanoic acids (Kim and Lee, 2016). Furthermore, cell lysis and programmed cell death has generally been believed to lead to specialized differentiation of biofilm bacteria (Kim and Lee, 2016) and the *P.*

aeruginosa detachment. This autolysis is controlled by the *cid/lrg* regulatory system, which is functionally and structurally related to Pf4 bacteriophage-holins that regulate host cell lysis by modulating the expression of murein hydrolase (Kim and Lee, 2016) and by the 2-heptyl-3-hydroxyl-4-quinolone quorum sensing system (D'Argenio et al., 2002). Curiously, Pf4 phage plays an important role in *P. aeruginosa* biofilm development and is involved in the formation of small colony variants and the virulence factor expression (Rice et al., 2009). Finally, *Candida* biofilms release single yeast cells from the upper hyphal layers which display enhanced adherence, filamentation and pathogenic capacities [203]. This process is assumed to be controlled by farnesol by inducing the production of several molecules implied in detachment and dispersal regulation, such as transcription factors NRG1 and PES1, [204], heat-shock protein Hsp90 [205] and cell wall protein Ywp1 [206] in *C. albicans*.

Host tissue integration and biofilm development: the race for the surface and a frustrated phagocytosis

The pathogenesis of biomaterial-associated infections is the result of the interaction between the microorganism, the implant and the host (McConoughey et al., 2014), which in itself adds a dimension of complexity to this type of infection.

From this complex interaction was suggested the term of "race for surface" by the orthopedic surgeon Anthony G. Gristina (Gristina, 1987). According to his postulate, after the conditioning layer on medical-device surface, the implant represents a ready surface for the colonization of both tissue cells and any microorganisms that were arrive on it during the device implantation. This way, a race between tissue cells and microorganism are established. If the race is won by tissue cells, the medical device surface will be covered by tissue surface and less vulnerable to microbial colonization. Otherwise, if the race is won by the microorganisms, the biomaterial surface will become rapidly covered by microbial biofilm and tissue cell functions will be hampered by microbial virulence factors and toxins. This last occurs most often because the biofilm development used to be faster than tissue integration of medical devices (Zhao et al., 2014).

The "race for the surface" has been demonstrated by several authors (Martinez-Perez et al., 2017; Perez-Tanoira et al., 2017a; Perez-Tanoira et al., 2017b; Pham et al., 2016; Subbiahdoss et al., 2009; Zhao et al., 2014), and their similarities and differences are summarized in the Table 6. All works published to date support the "race for surface", but some of them have used it for optimizing a flow model of cell-bacteria co-culture (Subbiahdoss et al., 2009), describing the best surface finishing for favoring the tissue integrations of metallic (Martinez-Perez et al., 2017; Zhao et al., 2014) and silicon (Pham et al., 2016) surfaces, or for postulating the *in-vitro* pre-incubation of implants as an anti-infective treatment before surgery (Perez-Tanoira et al., 2017a; Perez-Tanoira et al., 2017b).

Hematogenous acute-infections on biomaterials

Fibrous capsule, placed between the biomaterial surface and the surrounding tissue or interstitial medium, is a *locus minus resistentiae* characterized as an immunosuppressed fibro-inflammatory zone (Garcia-Gareta et al., 2016). This zone is a relatively inaccessible environment for the immune response, due to the absence of a normal blood supply in the peri-device tissue (McConoughey et al., 2014), which decreases the ability of neutrophils and monocytes/ macrophages, antibodies or certain antibiotics to correctly reach the biomaterial surface of the medical devices. In addition, at the level of innate cellular immunity, neutrophils are rapidly associated with the surface of the implanted devices (Kaplan et al., 1999). These prematurely activated neutrophils adhered on conditioning layer and loses the ability to activate in response to subsequent normal stimuli. In addition, the releasing of human neutrophilic peptides, under certain conditions, decreases the activity of other polymorphonuclear cells creating a hostile environment for the host immune system (Kaplan et al., 1999), where phagocytosis, granular enzyme levels and the ability to trigger respiratory burst decrease (Bingham et al., 2014). Therefore, the implant will be susceptible to infection not only during the perioperative period, but throughout its useful life (Zimmerli et al., 2004).

Table 6. Summary of studies which have evaluated "race for surface" postulate.

Bacterial species	Cell types	Substrata	Reference
Staphylococcus epidermidis	U-2 OS human bone osteosarcoma cells	Glass	(Subbiahdoss et al., 2009)
Streptococcus oralis, S. mitis and *S. salivariums* *Staphylococcus aureus*	HGF-1 gingival human fibroblasts	Titanium alloy Titanium-zirconium alloy Zirconia	(Zhao et al., 2014)
S. aureus *Pseudomonas aeruginosa*	COS-7 African green monkey kidney fibroblasts	Black silicon	(Pham et al., 2016)
S. aureus *S. epidermidis*	MC3T3-E1 mouse preosteoblasts	Ti-6AL-4V alloy	(Martinez-Perez et al., 2017)
S. aureus	SAOS-2 primary human osteogenic sarcoma cells hOB primary human osteoblasts	Ti deposited by sputtering Polydimethylsiloxane Polyethylene	(Perez-Tanoira et al., 2017b)
S. aureus	SAOS-2 primary human osteogenic sarcoma cells	Ti deposited by sputtering Polydimethylsiloxane Polyethylene	(Perez-Tanoira et al., 2017a)

Hematogenous acute infections related to medical devices (HAI-MD) are infections provoked by post-operative bacteremia from any source and that may lead to hematogenous seeding and infection of any implanted medical device (Vergidis and Patel, 2012). This bacteremia may hypothetically be symptomatic or asymptomatic and originated from dental, gastrointestinal, genitourinary or other invasive procedures or infection (Vergidis and Patel, 2012), but there are no enough evidences about it in the literature. The most studied HAI-MD are the

PJI where this type of infections represents between 20 and 40% of the total of PJI (Tande et al., 2016; Wang et al., 2017). In this hematogenous PJI, Gram-positive bacteria are the often the causative agents (mainly staphylococci and streptococci) (Rodriguez et al., 2010; Rui et al., 2016). Furthermore, it is known that *S. aureus* bacteremia results in a hematogenous PJI in the 30-40% of patients with a joint prosthesis (Sendi et al., 2011; Tande et al., 2016). Recently, it has been demonstrated using an *in-vivo* murine model that the hematogenous PJI are developed and established on well-vascularized implant area and that ClfA (clumping factor A) plays a crucial role on the development of *S. aureus* hematogenous PJI (Wang et al., 2017).

Conclusions and future trends

Biomaterial-related infections are a clear example of biofilm-related diseases. The current knowledge of these phenomenon allowed us to know several mechanisms involved in the different steps of biofilm development, not only in one organism, but in a broad spectrum of different species involved in clinical disease. However, many aspects are still unknown and the translation of these knowledge to clinical practice is still to come. Biofilm development is a well-known mechanism of antimicrobial resistance and further research must be aimed to increase the knowledge of biofilm development with focus on the development of new mechanisms that allow us to manage these infections in a more effective way than the presently available methods.

Web Resources

http://www.biofilm.montana.edu/ Center for Biofilm Engineering. Montana State University.

https://www.cdc.gov/ Centers for Disease Control and Prevention.

https://www.pasteur.fr/fr/institut-pasteur Institut Pasteur. France.

https://www.escmid.org/research_projects/study_groups/biofilms/ ESCMID Study Group for Biofilms (ESGB).

https://www.escmid.org/research_projects/study_groups/implant_infections/ ESCMID Study Group for Implant-Associated Infections (ESGIAI).

https://isim.ku.dk/ : Department of Immunology and Microbiology. University of Copenhagen.

https://medicine.osu.edu/mii/Pages/index.aspx Microbial Infection and Immunology. Ohio State University.

https://www.asm.org/ American Society for Microbiology.

References

Adam, E.C., Mitchell, B.S., Schumacher, D.U., Grant, G., and Schumacher, U. (1997). *Pseudomonas aeruginosa* II lectin stops human ciliary beating: therapeutic implications of fucose. American journal of respiratory and critical care medicine *155*, 2102-2104 DOI: 10.1164/ajrccm.155.6.9196121.

Aggarwal, V.K., Bakhshi, H., Ecker, N.U., Parvizi, J., Gehrke, T., and Kendoff, D. (2014). Organism profile in periprosthetic joint infection: pathogens differ at two arthroplasty infection referral centers in Europe and in the United States. The journal of knee surgery *27*, 399-406 DOI: 10.1055/s-0033-1364102.

Ahmer, B.M. (2004). Cell-to-cell signalling in *Escherichia coli* and *Salmonella enterica*. Mol Microbiol *52*, 933-945 DOI: 10.1111/j.1365-2958.2004.04054.x.

Alberici, I., Bayazit, A.K., Drozdz, D., Emre, S., Fischbach, M., Harambat, J., Jankauskiene, A., Litwin, M., Mir, S., Morello, W., *et al.* (2015). Pathogens causing urinary tract infections in infants: a European overview by the ESCAPE study group. European journal of pediatrics *174*, 783-790 DOI: 10.1007/s00431-014-2459-3.

Alem, M.A., Oteef, M.D., Flowers, T.H., and Douglas, L.J. (2006). Production of tyrosol by *Candida albicans* biofilms and its role in quorum sensing and biofilm development. Eukaryotic cell *5*, 1770-1779 DOI: 10.1128/EC.00219-06.

Alexander, E.H., Rivera, F.A., Marriott, I., Anguita, J., Bost, K.L., and Hudson, M.C. (2003). *Staphylococcus aureus* - induced tumor necrosis factor - related apoptosis - inducing ligand expression mediates apoptosis and caspase-8 activation in infected osteoblasts. BMC microbiology *3*, 5.

Alhede, M., Bjarnsholt, T., Givskov, M., and Alhede, M. (2014). *Pseudomonas aeruginosa* biofilms: mechanisms of immune evasion. Advances in applied microbiology *86*, 1-40 DOI: 10.1016/B978-0-12-800262-9.00001-9.

Anderson, J.M., Bonfield, T.L., and Ziats, N.P. (1990). Protein adsorption and cellular adhesion and activation on biomedical polymers. The International journal of artificial organs *13*, 375-382.

Anderson, J.M., Rodriguez, A., and Chang, D.T. (2008). Foreign body reaction to biomaterials. Seminars in immunology *20*, 86-100 DOI: 10.1016/j.smim.2007.11.004.

Araujo, D., Henriques, M., and Silva, S. (2017). Portrait of Candida Species Biofilm Regulatory Network Genes. Trends Microbiol *25*, 62-75 DOI: 10.1016/j.tim.2016.09.004.

Arciola, C.R., Alvi, F.I., An, Y.H., Campoccia, D., and Montanaro, L. (2005). Implant infection and infection resistant materials: a mini review. The International journal of artificial organs *28*, 1119-1125.

Arciola, C.R., Campoccia, D., and Montanaro, L. (2018). Implant infections: adhesion, biofilm formation and immune evasion. Nat Rev Microbiol *16*, 397-409 DOI: 10.1038/s41579-018-0019-y.

Baker, C.S., Eory, L.A., Yakhnin, H., Mercante, J., Romeo, T., and Babitzke, P. (2007). CsrA inhibits translation initiation of Escherichia coli hfq by binding to a single site overlapping the Shine-Dalgarno sequence. Journal of bacteriology *189*, 5472-5481 DOI: 10.1128/JB.00529-07.

Banas, J.A., and Vickerman, M.M. (2003). Glucan-binding proteins of the oral streptococci. Critical reviews in oral biology and medicine : an official publication of the American Association of Oral Biologists *14*, 89-99.

Bassler, B.L. (1999). How bacteria talk to each other: regulation of gene expression by quorum sensing. Current opinion in microbiology *2*, 582-587.

Bayles, K.W. (2007). The biological role of death and lysis in biofilm development. Nat Rev Microbiol *5*, 721-726 DOI: 10.1038/nrmicro1743.

Beg, A.M., Jones, M.N., Miller-Torbert, T., and Holt, R.G. (2002). Binding of *Streptococcus mutans* to extracellular matrix molecules and fibrinogen. Biochemical and biophysical research communications *298*, 75-79.

Benito, N., Franco, M., Ribera, A., Soriano, A., Rodriguez-Pardo, D., Sorli, L., Fresco, G., Fernandez-Sampedro, M., Dolores Del Toro, M., Guio, L., *et al.* (2016). Time trends in the aetiology of prosthetic joint infections: a multicentre cohort study. Clinical microbiology and infection : the official publication of the

European Society of Clinical Microbiology and Infectious Diseases *22*, 732 e731-738 DOI: S1198-743X(16)30128-8 [pii] 10.1016/j.cmi.2016.05.004.

Bianco, A., Capano, M.S., Mascaro, V., Pileggi, C., and Pavia, M. (2018). Prospective surveillance of healthcare-associated infections and patterns of antimicrobial resistance of pathogens in an Italian intensive care unit. Antimicrobial resistance and infection control *7*, 48 DOI: 10.1186/s13756-018-0337-x.

Bingham, J., Clarke, H., Spangehl, M., Schwartz, A., Beauchamp, C., and Goldberg, B. (2014). The alpha defensin-1 biomarker assay can be used to evaluate the potentially infected total joint arthroplasty. Clinical orthopaedics and related research *472*, 4006-4009 DOI: 10.1007/s11999-014-3900-7.

Bjarnsholt, T., Alhede, M., Eickhardt-Sorensen, S.R., Moser, C., Kuhl, M., Jensen, P.O., and Hoiby, N. (2013). The *in vivo* biofilm. Trends Microbiol *21*, 466-474 DOI: S0966-842X(13)00111-X [pii] 10.1016/j.tim.2013.06.002.

Bokarewa, M.I., Jin, T., and Tarkowski, A. (2006). *Staphylococcus aureus*: Staphylokinase. The international journal of biochemistry & cell biology *38*, 504-509 DOI: 10.1016/j.biocel.2005.07.005.

Boles, B.R., and Horswill, A.R. (2008). Agr-mediated dispersal of *Staphylococcus aureus* biofilms. PLoS pathogens *4*, e1000052 DOI: 10.1371/journal.ppat.1000052.

Bos, R., van der Mei, H.C., and Busscher, H.J. (1999). Physico-chemistry of initial microbial adhesive interactions--its mechanisms and methods for study. FEMS Microbiol Rev *23*, 179-230.

Bose, J.L., Lehman, M.K., Fey, P.D., and Bayles, K.W. (2012). Contribution of the *Staphylococcus aureus* Atl AM and GL murein hydrolase activities in cell division, autolysis, and biofilm formation. PloS one *7*, e42244 DOI: 10.1371/journal.pone.0042244.

Boyd, A., and Chakrabarty, A.M. (1995). *Pseudomonas aeruginosa* biofilms: role of the alginate exopolysaccharide. Journal of industrial microbiology *15*, 162-168.

Brinkmann, V., Reichard, U., Goosmann, C., Fauler, B., Uhlemann, Y., Weiss, D.S., Weinrauch, Y., and Zychlinsky, A. (2004). Neutrophil extracellular traps kill bacteria. Science *303*, 1532-1535 DOI: 10.1126/science.1092385.

Brown, E.J., Joiner, K.A., Cole, R.M., and Berger, M. (1983). Localization of complement component 3 on *Streptococcus pneumoniae*: anti-capsular antibody causes complement deposition on the pneumococcal capsule. Infection and immunity *39*, 403-409.

Bruzaud, J., b, Tarradec, J., Coudreused, A., Canette, A., Herrya, J-M.,b, Taffin de Givenchyc, E., Darmanin, T., Guittard, F., Guilbaud, M., Bellon-Fontainea, M-N. (2015). Flagella but not type IV pili are involved in the initial adhesion of *Pseudomonas aeruginosa* PAO1 to hydrophobic or superhydrophobic surfaces. Colloids and Surfaces B: Biointerfaces *131*, 59-66 DOI: 10.1016/j.colsurfb.2015.04.036.

Bui, L.M., Conlon, B.P., and Kidd, S.P. (2017). Antibiotic tolerance and the alternative lifestyles of *Staphylococcus aureus*. Essays in biochemistry *61*, 71-79 DOI: 10.1042/EBC20160061.

Campoccia, D., Montanaro, L., and Arciola, C.R. (2006). The significance of infection related to orthopedic devices and issues of antibiotic resistance. Biomaterials *27*, 2331-2339 DOI: 10.1016/j.biomaterials.2005.11.044.

Campoccia, D., Testoni, F., Ravaioli, S., Cangini, I., Maso, A., Speziale, P., Montanaro, L., Visai, L., and Arciola, C.R. (2016). Orthopedic implant infections: Incompetence of *Staphylococcus epidermidis*, *Staphylococcus lugdunensis*, and *Enterococcus faecalis* to invade osteoblasts. Journal of biomedical materials research Part A *104*, 788-801 DOI: 10.1002/jbm.a.35564.

Cardile, A.P., Sanchez, C.J., Jr., Samberg, M.E., Romano, D.R., Hardy, S.K., Wenke, J.C., Murray, C.K., and Akers, K.S. (2014). Human plasma enhances the expression of Staphylococcal microbial surface components recognizing adhesive matrix molecules promoting biofilm formation and increases antimicrobial tolerance In Vitro. BMC research notes *7*, 457 DOI: 10.1186/1756-0500-7-457.

Chen, H., Fujita, M., Feng, Q., Clardy, J., and Fink, G.R. (2004). Tyrosol is a quorum-sensing molecule in *Candida albicans*. Proceedings of the National Academy of Sciences of the United States of America *101*, 5048-5052 DOI: 10.1073/pnas.0401416101.

Chow, J.W., Thal, L.A., Perri, M.B., Vazquez, J.A., Donabedian, S.M., Clewell, D.B., and Zervos, M.J. (1993). Plasmid-associated hemolysin and aggregation substance production contribute to virulence in experimental enterococcal endocarditis. Antimicrobial agents and chemotherapy *37*, 2474-2477.

Chrzanowska, N.Z.-R., M. (2014). The impacts of aluminum and zirconium nano-oxides on planktonic and biofilm bacteria. Desalination and Water Treatment *52*, 3680-3689 DOI: 10.1080/19443994.2014.884528.

Costerton, J.W., Lewandowski, Z., Caldwell, D.E., Korber, D.R., and Lappin-Scott, H.M. (1995). Microbial biofilms. Annu Rev Microbiol *49*, 711-745 DOI: 10.1146/annurev.mi.49.100195.003431.

Cvitkovitch, D.G. (2001). Genetic competence and transformation in oral streptococci. Critical reviews in oral biology and medicine : an official publication of the American Association of Oral Biologists *12*, 217-243.

D'Argenio, D.A., Calfee, M.W., Rainey, P.B., and Pesci, E.C. (2002). Autolysis and autoaggregation in *Pseudomonas aeruginosa* colony morphology mutants. Journal of bacteriology *184*, 6481-6489.

Daddaoua, A., Fillet, S., Fernandez, M., Udaondo, Z., Krell, T., and Ramos, J.L. (2012). Genes for carbon metabolism and the ToxA virulence factor in *Pseudomonas aeruginosa* are regulated through molecular interactions of PtxR and PtxS. PloS one *7*, e39390 DOI: 10.1371/journal.pone.0039390.

Dakheel, K.H., Abdul Rahim, R., Neela, V.K., Al-Obaidi, J.R., Hun, T.G., and Yusoff, K. (2016). Methicillin-Resistant Staphylococcus aureus biofilms and their influence on bacterial adhesion and cohesion. BioMed research international *2016*, 4708425 DOI: 10.1155/2016/4708425.

Das, T., Sehar, S., and Manefield, M. (2013). The roles of extracellular DNA in the structural integrity of extracellular polymeric substance and bacterial biofilm development. Environmental microbiology reports *5*, 778-786 DOI: 10.1111/1758-2229.12085.

Das, T., Sharma, P.K., Krom, B.P., van der Mei, H.C., and Busscher, H.J. (2011). Role of eDNA on the adhesion forces between *Streptococcus mutans* and substratum surfaces: influence of ionic strength and substratum hydrophobicity. Langmuir : the ACS journal of surfaces and colloids *27*, 10113-10118 DOI: 10.1021/la202013m.

Dastgheyb, S.S., Villaruz, A.E., Le, K.Y., Tan, V.Y., Duong, A.C., Chatterjee, S.S., Cheung, G.Y., Joo, H.S., Hickok, N.J., and Otto, M. (2015). Role of phenol-soluble modulins in formation of *Staphylococcus aureus* biofilms in synovial fluid. Infection and immunity *83*, 2966-2975 DOI: 10.1128/IAI.00394-15.

Davey, M.E., Caiazza, N.C., and O'Toole, G.A. (2003). Rhamnolipid surfactant production affects biofilm architecture in *Pseudomonas aeruginosa* PAO1. Journal of bacteriology *185*, 1027-1036.

De Keersmaecker, S.C., Sonck, K., and Vanderleyden, J. (2006). Let LuxS speak up in AI-2 signaling. Trends Microbiol *14*, 114-119 DOI: 10.1016/j.tim.2006.01.003.

DeFrancesco, A.S., Masloboeva, N., Syed, A.K., DeLoughery, A., Bradshaw, N., Li, G.W., Gilmore, M.S., Walker, S., and Losick, R. (2017). Genome-wide screen for genes involved in eDNA release during biofilm formation by *Staphylococcus aureus*. Proceedings of the National Academy of Sciences of the United States of America *114*, E5969-E5978 DOI: 10.1073/pnas.1704544114.

Denning, G.M., Wollenweber, L.A., Railsback, M.A., Cox, C.D., Stoll, L.L., and Britigan, B.E. (1998). Pseudomonas pyocyanin increases interleukin-8 expression by human airway epithelial cells. Infection and immunity *66*, 5777-5784.

Deorukhkar, S.C., and Saini, S. (2016). Medical device-associated Candida infections in a rural tertiary care teaching hospital of India. Interdisciplinary perspectives on infectious diseases *2016*, 1854673 DOI: 10.1155/2016/1854673.

Deva, A.K., Adams, W.P., Jr., and Vickery, K. (2013). The role of bacterial biofilms in device-associated infection. Plastic and reconstructive surgery *132*, 1319-1328 DOI: 10.1097/PRS.0b013e3182a3c105.

Diekema, D.J., Pfaller, M.A., Schmitz, F.J., Smayevsky, J., Bell, J., Jones, R.N., Beach, M., and Group, S.P. (2001). Survey of infections due to Staphylococcus species: frequency of occurrence and antimicrobial susceptibility of isolates collected in the United States, Canada, Latin America, Europe, and the Western Pacific region for the SENTRY Antimicrobial Surveillance Program, 1997-1999. Clinical infectious diseases : an official publication of the Infectious Diseases Society of America *32 Suppl 2*, S114-132 DOI: 10.1086/320184.

Diggle, S.P. (2010). Microbial communication and virulence: lessons from evolutionary theory. Microbiology *156*, 3503-3512 DOI: mic.0.045179-0 [pii] 10.1099/mic.0.045179-0.

Diggle, S.P., Cornelis, P., Williams, P., and Camara, M. (2006). 4-quinolone signalling in *Pseudomonas aeruginosa*: old molecules, new perspectives. International journal of medical microbiology : IJMM *296*, 83-91 DOI: 10.1016/j.ijmm.2006.01.038.

Dinges, M.M., Orwin, P.M., and Schlievert, P.M. (2000). Exotoxins of *Staphylococcus aureus*. Clinical microbiology reviews *13*, 16-34, table of contents.

Donlan, R.M. (2001). Biofilms and device-associated infections. Emerg Infect Dis *7*, 277-281 DOI: 10.3201/eid0702.700277.

dos Santos, V., Brandalise, R.N., and Savaris, M. (2017). Biomaterials: Characteristics and properties. In Engineering of biomaterials (Cham: Springer International Publishing), pp. 5-15 DOI: 10.1007/978-3-319-58607-6_2.

Douglas, L.J. (2002). Medical importance of biofilms in Candida infections. Revista iberoamericana de micologia *19*, 139-143.

Dubin, G., Chmiel, D., Mak, P., Rakwalska, M., Rzychon, M., and Dubin, A. (2001). Molecular cloning and biochemical characterisation of proteases from *Staphylococcus epidermidis*. Biological chemistry *382*, 1575-1582 DOI: 10.1515/BC.2001.192.

Emody, L., Kerenyi, M., and Nagy, G. (2003). Virulence factors of uropathogenic *Escherichia coli*. International journal of antimicrobial agents *22 Suppl 2*, 29-33.

Ene, I.V., and Bennett, R.J. (2009). Hwp1 and related adhesins contribute to both mating and biofilm formation in *Candida albicans*. Eukaryotic cell *8*, 1909-1913 DOI: 10.1128/EC.00245-09.

Engelbert, M., Mylonakis, E., Ausubel, F.M., Calderwood, S.B., and Gilmore, M.S. (2004). Contribution of gelatinase, serine protease, and fsr to the pathogenesis of *Enterococcus faecalis* endophthalmitis. Infection and immunity *72*, 3628-3633 DOI: 10.1128/IAI.72.6.3628-3633.2004.

Erlandsen, S.L., Kristich, C.J., Dunny, G.M., and Wells, C.L. (2004). High-resolution visualization of the microbial glycocalyx with low-voltage scanning electron microscopy: dependence on cationic dyes. The journal of histochemistry and cytochemistry : official journal of the Histochemistry Society *52*, 1427-1435 DOI: 10.1369/jhc.4A6428.2004.

Fischer, C.R., Yoichi, M., Unno, H., and Tanji, Y. (2004). The coexistence of *Escherichia coli* serotype O157:H7 and its specific bacteriophage in continuous culture. FEMS microbiology letters *241*, 171-177 DOI: S0378-1097(04)00735-9 [pii]10.1016/j.femsle.2004.10.017.

Flemming, H.C., and Wingender, J. (2010). The biofilm matrix. Nat Rev Microbiol *8*, 623-633 DOI: nrmicro2415 [pii] 10.1038/nrmicro2415.

Flemming, H.C., Wingender, J., Szewzyk, U., Steinberg, P., Rice, S.A., and Kjelleberg, S. (2016). Biofilms: an emergent form of bacterial life. Nat Rev Microbiol *14*, 563-575 DOI: nrmicro.2016.94 [pii] 10.1038/nrmicro.2016.94.

Foster, S.J. (1995). Molecular characterization and functional analysis of the major autolysin of *Staphylococcus aureus* 8325/4. Journal of bacteriology *177*, 5723-5725.

Foster, T.J., Geoghegan, J.A., Ganesh, V.K., and Hook, M. (2014). Adhesion, invasion and evasion: the many functions of the surface proteins of *Staphylococcus aureus*. Nat Rev Microbiol *12*, 49-62 DOI: 10.1038/nrmicro3161.

Fraser, J.D., and Proft, T. (2008). The bacterial superantigen and superantigen-like proteins. Immunological reviews *225*, 226-243 DOI: 10.1111/j.1600-065X.2008.00681.x.

Friedlander, R.S., Vogel, N., and Aizenberg, J. (2015). Role of flagella in adhesion of *Escherichia coli* to abiotic surfaces. Langmuir : the ACS journal of surfaces and colloids *31*, 6137-6144 DOI: 10.1021/acs.langmuir.5b00815.

Frohner, I.E., Bourgeois, C., Yatsyk, K., Majer, O., and Kuchler, K. (2009). *Candida albicans* cell surface superoxide dismutases degrade host-derived reactive oxygen species to escape innate immune surveillance. Mol Microbiol *71*, 240-252 DOI: 10.1111/j.1365-2958.2008.06528.x.

Garcia-Contreras, R., Zhang, X.S., Kim, Y., and Wood, T.K. (2008). Protein translation and cell death: the role of rare tRNAs in biofilm formation and in activating dormant phage killer genes. PloS one *3*, e2394 DOI: 10.1371/journal.pone.0002394.

Garcia-Gareta, E., Davidson, C., Levin, A., Coathup, M.J., and Blunn, G.W. (2016). Biofilm formation in total hip arthroplasty: prevention and treatment. RSC Advances *6*, 80244-80261 DOI: 10.1039/c6ra09583f.

Garrett, T.R., Bhakoo, M., and Zhang, Z. (2008). Bacterial adhesion and biofilms on surfaces. Progress in Natural Science *18*, 1049-1056 DOI: https://doi.org/10.1016/j.pnsc.2008.04.001.

Geetha, M., Singh, A.K., Asokamani, R., and Gogia, A.K. (2009). Ti based biomaterials, the ultimate choice for orthopaedic implants – A review. Progress in Materials Science *54*, 397-425 DOI: https://doi.org/10.1016/j.pmatsci.2008.06.004.

Geiger, T., Francois, P., Liebeke, M., Fraunholz, M., Goerke, C., Krismer, B., Schrenzel, J., Lalk, M., and Wolz, C. (2012). The stringent response of *Staphylococcus aureus* and its impact on survival after phagocytosis through the induction of intracellular PSMs expression. PLoS pathogens *8*, e1003016 DOI: 10.1371/journal.ppat.1003016.

Gendreau, L., and Loewy, Z.G. (2011). Epidemiology and etiology of denture stomatitis. Journal of prosthodontics : official journal of the American College of Prosthodontists *20*, 251-260 DOI: 10.1111/j.1532-849X.2011.00698.x.

Geoghegan, J.A., Monk, I.R., O'Gara, J.P., and Foster, T.J. (2013). Subdomains N2N3 of fibronectin binding protein A mediate *Staphylococcus aureus* biofilm formation and adherence to fibrinogen using distinct mechanisms. Journal of bacteriology *195*, 2675-2683 DOI: 10.1128/JB.02128-12.

Ghafoor, A., Hay, I.D., and Rehm, B.H. (2011). Role of exopolysaccharides in *Pseudomonas aeruginosa* biofilm formation and architecture. Appl Environ Microbiol *77*, 5238-5246 DOI: 10.1128/AEM.00637-11.

Giltner, C.L., van Schaik, E.J., Audette, G.F., Kao, D., Hodges, R.S., Hassett, D.J., and Irvin, R.T. (2006). The *Pseudomonas aeruginosa* type IV pilin receptor binding domain functions as an adhesin for both biotic and abiotic surfaces. Mol Microbiol *59*, 1083-1096 DOI: MMI5002 [pii]10.1111/j.1365-2958.2005.05002.x.

Goni, F.M., and Ostolaza, H. (1998). *E. coli* alpha-hemolysin: a membrane-active protein toxin. Brazilian journal of medical and biological research = Revista brasileira de pesquisas medicas e biologicas *31*, 1019-1034.

Graille, M., Stura, E.A., Corper, A.L., Sutton, B.J., Taussig, M.J., Charbonnier, J.B., and Silverman, G.J. (2000). Crystal structure of a *Staphylococcus aureus* protein A domain complexed with the Fab fragment of a human IgM antibody: structural basis for recognition of B-cell receptors and superantigen activity. Proceedings of the National Academy of Sciences of the United States of America *97*, 5399-5404.

Gristina, A.G. (1987). Biomaterial-centered infection: microbial adhesion versus tissue integration. Science *237*, 1588-1595.

Gross, M., Cramton, S.E., Gotz, F., and Peschel, A. (2001). Key role of teichoic acid net charge in *Staphylococcus aureus* colonization of artificial surfaces. Infection and immunity *69*, 3423-3426 DOI: 10.1128/IAI.69.5.3423-3426.2001.

Guignot, J., Chaplais, C., Coconnier-Polter, M.H., and Servin, A.L. (2007). The secreted autotransporter toxin, Sat, functions as a virulence factor in Afa/Dr diffusely adhering *Escherichia coli* by promoting lesions in tight junction of polarized epithelial cells. Cellular microbiology *9*, 204-221 DOI: 10.1111/j.1462-5822.2006.00782.x.

Guzman, C.A., Pruzzo, C., Plate, M., Guardati, M.C., and Calegari, L. (1991). Serum dependent expression of *Enterococcus faecalis* adhesins involved in the colonization of heart cells. Microbial pathogenesis *11*, 399-409.

Hall-Stoodley, L., Costerton, J.W., and Stoodley, P. (2004). Bacterial biofilms: from the natural environment to infectious diseases. Nat Rev Microbiol *2*, 95-108 DOI: 10.1038/nrmicro821.

Hanawa, T. (2012). Research and development of metals for medical devices based on clinical needs. Sci Technol Adv Mater *13*, 064102 DOI: 10.1088/1468-6996/13/6/064102 TSTA11661098 [pii].

Handley, P.S., Harty, D.W., Wyatt, J.E., Brown, C.R., Doran, J.P., and Gibbs, A.C. (1987). A comparison of the adhesion, coaggregation and cell-surface hydrophobicity properties of fibrillar and fimbriate strains of *Streptococcus salivarius*. Journal of general microbiology *133*, 3207-3217 DOI: 10.1099/00221287-133-11-3207.

He, W., and Benson, R. (2014). 4 - Polymeric Biomaterials A2 - Modjarrad, Kayvon. In handbook of polymer applications in medicine and medical devices, S. Ebnesajjad, ed. (Oxford: William Andrew Publishing), pp. 55-76 DOI: https://doi.org/10.1016/B978-0-323-22805-3.00004-9.

Heilmann, C., Hussain, M., Peters, G., and Gotz, F. (1997). Evidence for autolysin-mediated primary attachment of *Staphylococcus epidermidis* to a polystyrene surface. Mol Microbiol *24*, 1013-1024.

Henriques, M., Azeredo, J., and Oliveira, R. (2006). Candida species adhesion to oral epithelium: factors involved and experimental methodology used. Critical reviews in microbiology *32*, 217-226 DOI: 10.1080/10408410601023524.

Hornby, J.M., Jensen, E.C., Lisec, A.D., Tasto, J.J., Jahnke, B., Shoemaker, R., Dussault, P., and Nickerson, K.W. (2001). Quorum sensing in the dimorphic fungus *Candida albicans* is mediated by farnesol. Appl Environ Microbiol *67*, 2982-2992 DOI: 10.1128/AEM.67.7.2982-2992.2001.

Huang, G., Xia, D., An, T., Ng, T.W., Yip, H.Y., Li, G., Zhao, H., and Wong, P.K. (2015). Dual roles of capsular extracellular polymeric substances in photocatalytic inactivation of *Escherichia coli*: Comparison of *E. coli* BW25113 and isogenic mutants. Appl Environ Microbiol *81*, 5174-5183 DOI: AEM. 00775-15 [pii] 10.1128/AEM.00775-15.

Huebner, J., and Goldmann, D.A. (1999). Coagulase-negative staphylococci: role as pathogens. Annual review of medicine *50*, 223-236 DOI: 10.1146/annurev.med.50.1.223.

Hufnagel, D.A., Depas, W.H., and Chapman, M.R. (2015). The biology of the *Escherichia coli* extracellular matrix. Microbiology spectrum *3* DOI: 10.1128/microbiolspec.MB-0014-2014.

Hung, C., Zhou, Y., Pinkner, J.S., Dodson, K.W., Crowley, J.R., Heuser, J., Chapman, M.R., Hadjifrangiskou, M., Henderson, J.P., and Hultgren, S.J. (2013). *Escherichia coli* biofilms have an organized and complex extracellular matrix structure. mBio *4*, e00645-00613 DOI: 10.1128/mBio.00645-13.

Iordanou, S., Middleton, N., Papathanassoglou, E., and Raftopoulos, V. (2017). Surveillance of device associated infections and mortality in a major intensive care unit in the Republic of Cyprus. BMC infectious diseases *17*, 607 DOI: 10.1186/s12879-017-2704-2.

Jackowski, J.T., Szepfalusi, Z., Wanner, D.A., Seybold, Z., Sielczak, M.W., Lauredo, I.T., Adams, T., Abraham, W.M., and Wanner, A. (1991). Effects of *P. aeruginosa*-derived bacterial products on tracheal ciliary function: role of O2 radicals. The American journal of physiology *260*, L61-67 DOI: 10.1152/ajplung.1991.260.2.L61.

Jackson, D.W., Suzuki, K., Oakford, L., Simecka, J.W., Hart, M.E., and Romeo, T. (2002). Biofilm formation and dispersal under the influence of the global regulator CsrA of Escherichia coli. Journal of bacteriology *184*, 290-301.

Jaffer, I.H., Fredenburgh, J.C., Hirsh, J., and Weitz, J.I. (2015). Medical device-induced thrombosis: what causes it and how can we prevent it? Journal of thrombosis and haemostasis : JTH *13 Suppl 1*, S72-81 DOI: 10.1111/jth.12961.

Jenkinson, H.F., and Demuth, D.R. (1997). Structure, function and immunogenicity of streptococcal antigen I/II polypeptides. Mol Microbiol *23*, 183-190.

Jennings, L.K., Storek, K.M., Ledvina, H.E., Coulon, C., Marmont, L.S., Sadovskaya, I., Secor, P.R., Tseng, B.S., Scian, M., Filloux, A., *et al.* (2015). Pel is a cationic exopolysaccharide that cross-links extracellular DNA in the *Pseudomonas aeruginosa* biofilm matrix. Proceedings of the National Academy of Sciences of the United States of America *112*, 11353-11358 DOI: 10.1073/pnas.1503058112.

Jensen, V., Lons, D., Zaoui, C., Bredenbruch, F., Meissner, A., Dieterich, G., Munch, R., and Haussler, S. (2006). RhlR expression in *Pseudomonas aeruginosa* is modulated by the Pseudomonas quinolone signal via PhoB-dependent and -independent pathways. Journal of bacteriology *188*, 8601-8606 DOI: 10.1128/JB.01378-06.

Joo, H.S., Chan, J.L., Cheung, G.Y., and Otto, M. (2010). Subinhibitory concentrations of protein synthesis-inhibiting antibiotics promote increased expression of the agr virulence regulator and production of phenol-soluble modulin cytolysins in community-associated methicillin-resistant *Staphylococcus aureus*. Antimicrobial agents and chemotherapy *54*, 4942-4944 DOI: 10.1128/AAC.00064-10.

Joo, H.S., and Otto, M. (2012). Molecular basis of in vivo biofilm formation by bacterial pathogens. Chemistry & biology *19*, 1503-1513 DOI: 10.1016/j.chembiol.2012.10.022.

Jung, C.J., Hsu, R.B., Shun, C.T., Hsu, C.C., and Chia, J.S. (2017). AtlA mediates extracellular DNA Release, which contributes to *Streptococcus mutans* biofilm formation in an experimental rat model of infective endocarditis. Infection and immunity *85* DOI: 10.1128/IAI.00252-17.

Kalinka, J., Hachmeister, M., Geraci, J., Sordelli, D., Hansen, U., Niemann, S., Oetermann, S., Peters, G., Loffler, B., and Tuchscherr, L. (2014). *Staphylococcus aureus* isolates from chronic osteomyelitis are characterized by high host cell invasion and intracellular adaptation, but still induce inflammation. International journal of medical microbiology : IJMM *304*, 1038-1049 DOI: 10.1016/j.ijmm.2014.07.013.

Kanmani, P., Taik Lim, S. (2013). Synthesis and structural characterization of silver nanoparticles using bacterialexopolysaccharide and its antimicrobial activity against food and multidrugresistant pathogens. Process Biochemistry *48*, 1099-1106 DOI: 10.1016/j.procbio.2013.05.011.

Kaplan, J.B., and Fine, D.H. (2002). Biofilm dispersal of *Neisseria subflava* and other phylogenetically diverse oral bacteria. Appl Environ Microbiol *68* 4943-4950.

Kaplan, S.S., Heine, R.P., and Simmons, R.L. (1999). Defensins impair phagocytic killing by neutrophils in biomaterial-related infection. Infection and immunity *67*, 1640-1645.

Karchmer, A.W., and Longworth, D.L. (2002). Infections of intracardiac devices. Infectious disease clinics of North America *16*, 477-505, xii.

Kelly, C., Evans, P., Bergmeier, L., Lee, S.F., Progulske-Fox, A., Harris, A.C., Aitken, A., Bleiweis, A.S., and Lehner, T. (1989). Sequence analysis of the cloned streptococcal cell surface antigen I/II. FEBS letters *258*, 127-132.

Kessler, E., Safrin, M., Olson, J.C., and Ohman, D.E. (1993). Secreted LasA of *Pseudomonas aeruginosa* is a staphylolytic protease. The Journal of biological chemistry *268*, 7503-7508.

Kim, S.K., and Lee, J.H. (2016). Biofilm dispersion in *Pseudomonas aeruginosa*. Journal of microbiology *54*, 71-85 DOI: 10.1007/s12275-016-5528-7.

Kinnby, B., Booth, N.A., and Svensater, G. (2008). Plasminogen binding by oral streptococci from dental plaque and inflammatory lesions. Microbiology *154*, 924-931 DOI: 10.1099/mic.0.2007/013235-0.

Kocianova, S., Vuong, C., Yao, Y., Voyich, J.M., Fischer, E.R., DeLeo, F.R., and Otto, M. (2005). Key role of poly-gamma-DL-glutamic acid in immune evasion and virulence of *Staphylococcus epidermidis*. The Journal of clinical investigation *115*, 688-694 DOI: 10.1172/JCI23523.

Kojic, E.M., and Darouiche, R.O. (2004). Candida infections of medical devices. Clinical microbiology reviews *17*, 255-267.

Kruppa, M. (2009). Quorum sensing and *Candida albicans*. Mycoses *52*, 1-10 DOI: 10.1111/j.1439-0507.2008.01626.x.

Kuhn, D.M., Chandra, J., Mukherjee, P.K., and Ghannoum, M.A. (2002). Comparison of biofilms formed by *Candida albicans* and *Candida parapsilosis* on bioprosthetic surfaces. Infection and immunity *70*, 878-888.

Kuramitsu, H.K. (1993). Virulence factors of mutans streptococci: role of molecular genetics. Critical reviews in oral biology and medicine : an official publication of the American Association of Oral Biologists *4*, 159-176.

Kuroda, M., Ito, R., Tanaka, Y., Yao, M., Matoba, K., Saito, S., Tanaka, I., and Ohta, T. (2008). *Staphylococcus aureus* surface protein SasG contributes to intercellular autoaggregation of *Staphylococcus aureus*. Biochemical and biophysical research communications *377*, 1102-1106 DOI: 10.1016/j.bbrc.2008.10.134.

Laarman, A.J., Bardoel, B.W., Ruyken, M., Fernie, J., Milder, F.J., van Strijp, J.A., and Rooijakkers, S.H. (2012). *Pseudomonas aeruginosa* alkaline protease blocks complement activation via the classical and lectin pathways. Journal of immunology *188*, 386-393 DOI: 10.4049/jimmunol.1102162.

Laarman, A.J., Ruyken, M., Malone, C.L., van Strijp, J.A., Horswill, A.R., and Rooijakkers, S.H. (2011). *Staphylococcus aureus* metalloprotease aureolysin cleaves complement C3 to mediate immune evasion. Journal of immunology *186*, 6445-6453 DOI: 10.4049/jimmunol.1002948.

Lai, Y., Villaruz, A.E., Li, M., Cha, D.J., Sturdevant, D.E., and Otto, M. (2007). The human anionic antimicrobial peptide dermcidin induces proteolytic defence mechanisms in staphylococci. Mol Microbiol *63*, 497-506 DOI: 10.1111/j.1365-2958.2006.05540.x.

Latasa, C., Solano, C., Penades, J.R., and Lasa, I. (2006). Biofilm-associated proteins. Comptes rendus biologies *329*, 849-857 DOI: 10.1016/j.crvi.2006.07.008.

Latifi, A., Foglino, M., Tanaka, K., Williams, P., and Lazdunski, A. (1996). A hierarchical quorum-sensing cascade in *Pseudomonas aeruginosa* links the transcriptional activators LasR and RhlR (VsmR) to expression of the stationary-phase sigma factor RpoS. Mol Microbiol *21*, 1137-1146.

Lattif, A.A., Mukherjee, P.K., Chandra, J., Swindell, K., Lockhart, S.R., Diekema, D.J., Pfaller, M.A., and Ghannoum, M.A. (2010). Characterization of biofilms formed by *Candida parapsilosis*, *C. metapsilosis*, and *C. orthopsilosis*. International journal of medical microbiology : IJMM *300*, 265-270 DOI: 10.1016/j.ijmm.2009.09.001.

Lau, G.W., Haataja, S., Lonetto, M., Kensit, S.E., Marra, A., Bryant, A.P., McDevitt, D., Morrison, D.A., and Holden, D.W. (2001). A functional genomic analysis of type 3 *Streptococcus pneumoniae* virulence. Mol Microbiol *40*, 555-571.

Lau, G.W., Ran, H., Kong, F., Hassett, D.J., and Mavrodi, D. (2004). *Pseudomonas aeruginosa* pyocyanin is critical for lung infection in mice. Infection and immunity *72*, 4275-4278 DOI: 10.1128/IAI.72.7.4275-4278.2004.

Lauer, P., Rinaudo, C.D., Soriani, M., Margarit, I., Maione, D., Rosini, R., Taddei, A.R., Mora, M., Rappuoli, R., Grandi, G., *et al.* (2005). Genome analysis reveals pili in Group B Streptococcus. Science *309*, 105 DOI: 10.1126/science.1111563.

Laverty, G., Gorman, S.P., and Gilmore, B.F. (2014). Biomolecular Mechanisms of *Pseudomonas aeruginosa* and *Escherichia coli* Biofilm Formation. Pathogens *3*, 596-632 DOI: pathogens3030596 [pii] 10.3390/pathogens3030596.

Le, K.Y., Dastgheyb, S., Ho, T.V., and Otto, M. (2014). Molecular determinants of staphylococcal biofilm dispersal and structuring. Frontiers in cellular and infection microbiology *4*, 167 DOI: 10.3389/fcimb.2014.00167.

Lee, J., and Zhang, L. (2015). The hierarchy quorum sensing network in *Pseudomonas aeruginosa*. Protein & cell *6*, 26-41 DOI: 10.1007/s13238-014-0100-x.

Lee, K.H., Maiden, M.F., Tanner, A.C., and Weber, H.P. (1999). Microbiota of successful osseointegrated dental implants. Journal of periodontology *70*, 131-138 DOI: 10.1902/jop.1999.70.2.131.

Lee, M.S., and Morrison, D.A. (1999). Identification of a new regulator in *Streptococcus pneumoniae* linking quorum sensing to competence for genetic transformation. Journal of bacteriology *181*, 5004-5016.

Lee, W.G., Shin, J.H., Uh, Y., Kang, M.G., Kim, S.H., Park, K.H., and Jang, H.C. (2011). First three reported cases of nosocomial fungemia caused by *Candida auris*. Journal of clinical microbiology *49*, 3139-3142 DOI: 10.1128/JCM.00319-11.

Lequette, Y., and Greenberg, E.P. (2005). Timing and localization of rhamnolipid synthesis gene expression in *Pseudomonas aeruginosa* biofilms. Journal of bacteriology *187*, 37-44 DOI: 10.1128/JB.187.1.37-44.2005.

Li, F., and Palecek, S.P. (2008). Distinct domains of the *Candida albicans* adhesin Eap1p mediate cell-cell and cell-substrate interactions. Microbiology *154*, 1193-1203 DOI: 10.1099/mic.0.2007/013789-0.

Li, M., Lai, Y., Villaruz, A.E., Cha, D.J., Sturdevant, D.E., and Otto, M. (2007). Gram-positive three-component antimicrobial peptide-sensing system. Proceedings of the National Academy of Sciences of the United States of America *104*, 9469-9474 DOI: 10.1073/pnas.0702159104.

Lievin-Le Moal, V., Comenge, Y., Ruby, V., Amsellem, R., Nicolas, V., and Servin, A.L. (2011). Secreted autotransporter toxin (Sat) triggers autophagy in epithelial cells that relies on cell detachment. Cellular microbiology *13*, 992-1013 DOI: 10.1111/j.1462-5822.2011.01595.x.

Lin, M.H., Shu, J.C., Lin, L.P., Chong, K.Y., Cheng, Y.W., Du, J.F., and Liu, S.T. (2015). Elucidating the crucial role of poly N-acetylglucosamine from *Staphylococcus aureus* in cellular adhesion and pathogenesis. PloS one *10*, e0124216 DOI: 10.1371/journal.pone.0124216.

Lin, Y.C., and Peterson, M.L. (2010). New insights into the prevention of staphylococcal infections and toxic shock syndrome. Expert review of clinical pharmacology *3*, 753-767 DOI: 10.1586/ecp.10.121.

Liu, J., Sun, L., Liu, W., Guo, L., Liu, Z., Wei, X., and Ling, J. (2017). A Nuclease from *Streptococcus mutans* Facilitates Biofilm Dispersal and Escape from Killing by Neutrophil Extracellular Traps. Frontiers in cellular and infection microbiology *7*, 97 DOI: 10.3389/fcimb.2017.00097.

Loesche, W.J. (1996). Microbiology of Dental Decay and Periodontal Disease. In Medical Microbiology, th, and S. Baron, eds. (Galveston (TX)).

Long, M., and Rack, H.J. (1998). Titanium alloys in total joint replacement--a materials science perspective. Biomaterials *19*, 1621-1639 DOI: S0142-9612(97)00146-4 [pii].

Lorenz, M.C., Bender, J.A., and Fink, G.R. (2004). Transcriptional response of *Candida albicans* upon internalization by macrophages. Eukaryotic cell *3*, 1076-1087 DOI: 10.1128/EC.3.5.1076-1087.2004.

Lowe, A.M., Lambert, P.A., and Smith, A.W. (1995). Cloning of an *Enterococcus faecalis* endocarditis antigen: homology with adhesins from some oral streptococci. Infection and immunity *63*, 703-706.

Ludwig, A., von Rhein, C., Bauer, S., Huttinger, C., and Goebel, W. (2004). Molecular analysis of cytolysin A (ClyA) in pathogenic *Escherichia coli* strains. Journal of bacteriology *186*, 5311-5320 DOI: 10.1128/JB.186.16.5311-5320.2004.

Ma, L., Lu, H., Sprinkle, A., Parsek, M.R., and Wozniak, D.J. (2007). Pseudomonas aeruginosa Psl is a galactose- and mannose-rich exopolysaccharide. Journal of bacteriology *189*, 8353-8356 DOI: 10.1128/JB.00620-07.

Maali, Y., Martins-Simoes, P., Valour, F., Bouvard, D., Rasigade, J.P., Bes, M., Haenni, M., Ferry, T., Laurent, F., and Trouillet-Assant, S. (2016). Pathophysiological Mechanisms of Staphylococcus Non-aureus Bone and Joint Infection: Interspecies Homogeneity and Specific Behavior of *S. pseudintermedius*. Frontiers in microbiology *7*, 1063 DOI: 10.3389/fmicb.2016.01063.

Mack, D., Fischer, W., Krokotsch, A., Leopold, K., Hartmann, R., Egge, H., and Laufs, R. (1996a). The intercellular adhesin involved in biofilm accumulation of *Staphylococcus epidermidis* is a linear beta-1,6-linked glucosaminoglycan: purification and structural analysis. Journal of bacteriology *178*, 175-183.

Mack, D., Haeder, M., Siemssen, N., and Laufs, R. (1996b). Association of biofilm production of coagulase-negative staphylococci with expression of a specific polysaccharide intercellular adhesin. The Journal of infectious diseases *174*, 881-884.

Mack, D., Rohde, H., Dobinsky, S., Riedewald, J., Nedelmann, M., Knobloch, J.K., Elsner, H.A., and Feucht, H.H. (2000). Identification of three essential regulatory gene loci governing expression of *Staphylococcus epidermidis* polysaccharide intercellular adhesin and biofilm formation. Infection and immunity *68*, 3799-3807.

Maisey, H.C., Quach, D., Hensler, M.E., Liu, G.Y., Gallo, R.L., Nizet, V., and Doran, K.S. (2008). A group B streptococcal pilus protein promotes phagocyte resistance and systemic virulence. FASEB journal : official publication of the Federation of American Societies for Experimental Biology *22*, 1715-1724 DOI: 10.1096/fj.07-093963.

Malekani, J.S., B.; Gu, Y. T.; Schuetz, M.; Yarlagadda, P. K. (2012). Orthopedic bone plates : evolution in structure implementation technique and biomaterial. GSTF Journal of Engineering Technology (JET) *1*, 135-140 DOI: 10.5176/2251-3701_1.1.23.

Mann, E.E., Rice, K.C., Boles, B.R., Endres, J.L., Ranjit, D., Chandramohan, L., Tsang, L.H., Smeltzer, M.S., Horswill, A.R., and Bayles, K.W. (2009). Modulation of eDNA release and degradation affects *Staphylococcus aureus* biofilm maturation. PloS one *4*, e5822 DOI: 10.1371/journal.pone.0005822.

Martinez-Perez, M., Perez-Jorge, C., Lozano, D., Portal-Nunez, S., Perez-Tanoira, R., Conde, A., Arenas, M.A., Hernandez-Lopez, J.M., de Damborenea, J.J., Gomez-Barrena, E*., et al.* (2017). Evaluation of bacterial adherence of clinical isolates of Staphylococcus sp. using a competitive model: An in vitro approach to the "race for the surface" theory. Bone & joint research *6*, 315-322 DOI: 10.1302/2046-3758.65.BJR-2016-0226.R2.

McCann, M.T., Gilmore, B.F., and Gorman, S.P. (2008). *Staphylococcus epidermidis* device-related infections: pathogenesis and clinical management. The Journal of pharmacy and pharmacology *60*, 1551-1571 DOI: 10.1211/jpp/60.12.0001.

McConoughey, S.J., Howlin, R., Granger, J.F., Manring, M.M., Calhoun, J.H., Shirtliff, M., Kathju, S., and Stoodley, P. (2014). Biofilms in periprosthetic orthopedic infections. Future Microbiol *9*, 987-1007 DOI: 10.2217/fmb.14.64.

McCourt, J., O'Halloran, D.P., McCarthy, H., O'Gara, J.P., and Geoghegan, J.A. (2014). Fibronectin-binding proteins are required for biofilm formation by community-associated methicillin-resistant *Staphylococcus aureus* strain LAC. FEMS microbiology letters *353*, 157-164 DOI: 10.1111/1574-6968.12424.

Melzer, M., and Welch, C. (2013). Outcomes in UK patients with hospital-acquired bacteraemia and the risk of catheter-associated urinary tract infections. Postgraduate medical journal *89*, 329-334 DOI: 10.1136/postgradmedj-2012-131393.

Merritt, J., Qi, F., Goodman, S.D., Anderson, M.H., and Shi, W. (2003). Mutation of luxS affects biofilm formation in *Streptococcus mutans*. Infection and immunity *71*, 1972-1979.

Monod, M., and Borg-von, Z.M. (2002). Secreted aspartic proteases as virulence factors of Candida species. Biological chemistry *383*, 1087-1093 DOI: 10.1515/BC.2002.117.

Monroe, D. (2007). Looking for chinks in the armor of bacterial biofilms. PLoS Biol *5*, e307 DOI: 07-PLBI-F-2868 [pii] 10.1371/journal.pbio.0050307.

Mukherjee, P.K., Chandra, J., Kuhn, D.M., and Ghannoum, M.A. (2003). Mechanism of fluconazole resistance in *Candida albicans* biofilms: phase-specific role of efflux pumps and membrane sterols. Infection and immunity *71*, 4333-4340.

Murillo, L.A., Newport, G., Lan, C.Y., Habelitz, S., Dungan, J., and Agabian, N.M. (2005). Genome-wide transcription profiling of the early phase of biofilm formation by *Candida albicans*. Eukaryotic cell *4*, 1562-1573 DOI: 10.1128/EC.4.9.1562-1573.2005.

Nadell, C.D., Xavier, J.B., and Foster, K.R. (2009). The sociobiology of biofilms. FEMS Microbiol Rev *33*, 206-224 DOI: FMR150 [pii] 10.1111/j.1574-6976.2008.00150.x.

Nallapareddy, S.R., Singh, K.V., Sillanpaa, J., Garsin, D.A., Hook, M., Erlandsen, S.L., and Murray, B.E. (2006). Endocarditis and biofilm-associated pili of *Enterococcus faecalis*. The Journal of clinical investigation *116*, 2799-2807 DOI: 10.1172/JCI29021.

Navarathna, D.H., Hornby, J.M., Krishnan, N., Parkhurst, A., Duhamel, G.E., and Nickerson, K.W. (2007). Effect of farnesol on a mouse model of systemic candidiasis, determined by use of a DPP3 knockout mutant of *Candida albicans*. Infection and immunity *75*, 1609-1618 DOI: 10.1128/IAI.01182-06.

Nett, J.E., Crawford, K., Marchillo, K., and Andes, D.R. (2010). Role of Fks1p and matrix glucan in *Candida albicans* biofilm resistance to an echinocandin, pyrimidine, and polyene. Antimicrobial agents and chemotherapy *54*, 3505-3508 DOI: 10.1128/AAC.00227-10.

Neut, D., van der Mei, H.C., Bulstra, S.K., and Busscher, H.J. (2007). The role of small-colony variants in failure to diagnose and treat biofilm infections in orthopedics. Acta orthopaedica *78*, 299-308 DOI: 10.1080/17453670710013843.

Nguyen, V.T., Chia, T.W., Turner, M.S., Fegan, N., and Dykes, G.A. (2011). Quantification of acid-base interactions based on contact angle measurement allows XDLVO predictions to attachment of *Campylobacter jejuni* but not *Salmonella*. Journal of microbiological methods *86*, 89-96 DOI: 10.1016/j.mimet.2011.04.005.

Nicolle, L.E. (2014). Catheter associated urinary tract infections. Antimicrobial resistance and infection control *3*, 23 DOI: 10.1186/2047-2994-3-23.

Nielsen, X.C., Chen, M., Hellesoe, A.M., Jeppesen, P.B., Gyldenlykke, J., Tvede, M., and Andersen, L.P. (2012). Etiology and epidemiology of catheter related bloodstream infections in patients receiving home parenteral nutrition in a gastromedical center at a tertiary hospital in denmark. The open microbiology journal *6*, 98-101 DOI: 10.2174/1874285801206010098.

Nobbs, A.H., Lamont, R.J., and Jenkinson, H.F. (2009). Streptococcus adherence and colonization. Microbiology and molecular biology reviews : MMBR *73*, 407-450, Table of Contents DOI: 10.1128/MMBR.00014-09.

Nobile, C.J., and Mitchell, A.P. (2005). Regulation of cell-surface genes and biofilm formation by the *C. albicans* transcription factor Bcr1p. Current biology : CB *15*, 1150-1155 DOI: 10.1016/j.cub.2005.05.047.

Nobile, C.J., Nett, J.E., Andes, D.R., and Mitchell, A.P. (2006). Function of *Candida albicans* adhesin Hwp1 in biofilm formation. Eukaryotic cell *5*, 1604-1610 DOI: 10.1128/EC.00194-06.

Nobile, C.J., Schneider, H.A., Nett, J.E., Sheppard, D.C., Filler, S.G., Andes, D.R., and Mitchell, A.P. (2008). Complementary adhesin function in *C. albicans* biofilm formation. Current biology : CB *18*, 1017-1024 DOI: 10.1016/j.cub.2008.06.034.

Novick, R.P. (2003). Autoinduction and signal transduction in the regulation of staphylococcal virulence. Mol Microbiol *48*, 1429-1449.

Nuri, R., Shprung, T., and Shai, Y. (2015). Defensive remodeling: How bacterial surface properties and biofilm formation promote resistance to antimicrobial peptides. Biochimica et biophysica acta *1848*, 3089-3100 DOI: 10.1016/j.bbamem.2015.05.022.

Nuss, K.M., and von Rechenberg, B. (2008). Biocompatibility issues with modern implants in bone - a review for clinical orthopedics. The open orthopaedics journal *2*, 66-78 DOI: 10.2174/1874325000802010066.

O'Gara, J.P. (2007). ica and beyond: biofilm mechanisms and regulation in *Staphylococcus epidermidis* and *Staphylococcus aureus*. FEMS microbiology letters *270*, 179-188 DOI: 10.1111/j.1574-6968.2007.00688.x.

O'Grady N, P., Alexander, M., Dellinger, E.P., Gerberding, J.L., Heard, S.O., Maki, D.G., Masur, H., McCormick, R.D., Mermel, L.A., Pearson, M.L., *et al.* (2002). Guidelines for the prevention of intravascular catheter-related infections. American journal of infection control *30*, 476-489.

O'Neill, E., Pozzi, C., Houston, P., Humphreys, H., Robinson, D.A., Loughman, A., Foster, T.J., and O'Gara, J.P. (2008). A novel *Staphylococcus aureus* biofilm phenotype mediated by the fibronectin-binding proteins, FnBPA and FnBPB. Journal of bacteriology *190*, 3835-3850 DOI: 10.1128/JB.00167-08.

Obadia, B., Lacour, S., Doublet, P., Baubichon-Cortay, H., Cozzone, A.J., and Grangeasse, C. (2007). Influence of tyrosine-kinase Wzc activity on colanic acid production in *Escherichia coli* K12 cells. Journal of molecular biology *367*, 42-53 DOI: 10.1016/j.jmb.2006.12.048.

Ochsner, U.A., and Reiser, J. (1995). Autoinducer-mediated regulation of rhamnolipid biosurfactant synthesis in *Pseudomonas aeruginosa*. Proceedings of the National Academy of Sciences of the United States of America *92*, 6424-6428.

Oder, M., Arlic, M., Bohinc, K., and Fink, R. (2018). *Escherichia coli* biofilm formation and dispersion under hydrodynamic conditions on metal surfaces. International journal of environmental health research *28*, 55-63 DOI: 10.1080/09603123.2017.1415309.

Otto, M. (2001). *Staphylococcus aureus* and *Staphylococcus epidermidis* peptide pheromones produced by the accessory gene regulator agr system. Peptides *22*, 1603-1608.

Otto, M. (2009). *Staphylococcus epidermidis*--the 'accidental' pathogen. Nat Rev Microbiol *7*, 555-567 DOI: 10.1038/nrmicro2182.

Otto, M. (2013). Staphylococcal infections: mechanisms of biofilm maturation and detachment as critical determinants of pathogenicity. Annual review of medicine *64*, 175-188 DOI: 10.1146/annurev-med-042711-140023.

Pannanusorn, S., Fernandez, V., and Romling, U. (2013). Prevalence of biofilm formation in clinical isolates of Candida species causing bloodstream infection. Mycoses *56*, 264-272 DOI: 10.1111/myc.12014.

Park, P.W., Pier, G.B., Preston, M.J., Goldberger, O., Fitzgerald, M.L., and Bernfield, M. (2000). Syndecan-1 shedding is enhanced by LasA, a secreted virulence factor of *Pseudomonas aeruginosa*. The Journal of biological chemistry *275*, 3057-3064.

Passador, L., Cook, J.M., Gambello, M.J., Rust, L., and Iglewski, B.H. (1993). Expression of *Pseudomonas aeruginosa* virulence genes requires cell-to-cell communication. Science *260*, 1127-1130.

Patel, F., Goswami, P. & Khara, R. (2016). Detection of biofilm formation in device associated clinical bacterial isolates in cancer patients. Sri Lankan Journal of Infectious Diseases *6*, 8 DOI: http://doi.org/10.4038/sljid.v6i1.8086.

Patti, J.M., and Hook, M. (1994). Microbial adhesins recognizing extracellular matrix macromolecules. Current opinion in cell biology *6*, 752-758.

Perez-Tanoira, R., Aarnisalo, A.A., Eklund, K.K., Han, X., Soininen, A., Tiainen, V.M., Esteban, J., and Kinnari, T.J. (2017a). Prevention of Biomaterial Infection by Pre-Operative Incubation with Human Cells. Surgical infections *18*, 336-344 DOI: 10.1089/sur.2016.263.

Perez-Tanoira, R., Han, X., Soininen, A., Aarnisalo, A.A., Tiainen, V.M., Eklund, K.K., Esteban, J., and Kinnari, T.J. (2017b). Competitive colonization of prosthetic surfaces by staphylococcus aureus and human cells. Journal of biomedical materials research Part A *105*, 62-72 DOI: 10.1002/jbm.a.35863.

Pham, V.T., Truong, V.K., Orlowska, A., Ghanaati, S., Barbeck, M., Booms, P., Fulcher, A.J., Bhadra, C.M., Buividas, R., Baulin, V., *et al.* (2016). "Race for the Surface": Eukaryotic cells can win. ACS applied materials & interfaces *8*, 22025-22031 DOI: 10.1021/acsami.6b06415.

Pierce, C.G., Vila, T., Romo, J.A., Montelongo-Jauregui, D., Wall, G., Ramasubramanian, A., and Lopez-Ribot, J.L. (2017). The Candida albicans biofilm matrix: composition, structure and function. Journal of fungi *3* DOI: 10.3390/jof3010014.

Proctor, R.A., von Eiff, C., Kahl, B.C., Becker, K., McNamara, P., Herrmann, M., and Peters, G. (2006). Small colony variants: a pathogenic form of bacteria that facilitates persistent and recurrent infections. Nat Rev Microbiol *4*, 295-305 DOI: 10.1038/nrmicro1384.

Purevdorj-Gage, B., Costerton, W.J., and Stoodley, P. (2005). Phenotypic differentiation and seeding dispersal in non-mucoid and mucoid *Pseudomonas aeruginosa* biofilms. Microbiology *151*, 1569-1576 DOI: 10.1099/mic.0.27536-0.

Qin, X., Singh, K.V., Weinstock, G.M., and Murray, B.E. (2000). Effects of *Enterococcus faecalis* fsr genes on production of gelatinase and a serine protease and virulence. Infection and immunity *68*, 2579-2586.

Rakita, R.M., Vanek, N.N., Jacques-Palaz, K., Mee, M., Mariscalco, M.M., Dunny, G.M., Snuggs, M., Van Winkle, W.B., and Simon, S.I. (1999). *Enterococcus faecalis* bearing aggregation substance is resistant to killing by human neutrophils despite phagocytosis and neutrophil activation. Infection and immunity *67*, 6067-6075.

Ramage, G., Bachmann, S., Patterson, T.F., Wickes, B.L., and Lopez-Ribot, J.L. (2002a). Investigation of multidrug efflux pumps in relation to fluconazole resistance in *Candida albicans* biofilms. The Journal of antimicrobial chemotherapy *49*, 973-980.

Ramage, G., Martinez, J.P., and Lopez-Ribot, J.L. (2006). Candida biofilms on implanted biomaterials: a clinically significant problem. FEMS yeast research *6*, 979-986 DOI: 10.1111/j.1567-1364.2006.00117.x.

Ramage, G., Saville, S.P., Wickes, B.L., and Lopez-Ribot, J.L. (2002b) Inhibition of *Candida albicans* biofilm formation by farnesol, a quorum-sensing molecule. Appl Environ Microbiol *68*, 5459-5463.

Rice, K.C., Mann, E.E., Endres, J.L., Weiss, E.C., Cassat, J.E., Smeltzer, M.S. and Bayles, K.W. (2007). The cidA murein hydrolase regulator contributes to DNA release and biofilm development in *Staphylococcus aureus*. Proceedings of the National Academy of Sciences of the United States of America *104* 8113-8118 DOI: 10.1073/pnas.0610226104.

Rice, S.A., Tan, C.H., Mikkelsen, P.J., Kung, V., Woo, J., Tay, M., Hauser, A. McDougald, D., Webb, J.S., and Kjelleberg, S. (2009). The biofilm life cycle and virulence of *Pseudomonas aeruginosa* are dependent on a filamentous prophage. The ISME journal *3*, 271-282 DOI: 10.1038/ismej.2008.109.

Roberts, A.P., and Kreth, J. (2014). The impact of horizontal gene transfer on the adaptive ability of the human oral microbiome. Frontiers in cellular and infection microbiology *4*, 124 DOI: 10.3389/fcimb.2014.00124.

Rochford, E.T., Richards, R.G., and Moriarty, T.F. (2012). Influence of material on the development of device-associated infections. Clinical microbiology and infection : the official publication of the European Society of Clinical Microbiology and Infectious Diseases *18*, 1162-1167 DOI: 10.1111/j.1469-0691.2012.04002.x.

Rodriguez, D., Pigrau, C., Euba, G., Cobo, J., Garcia-Lechuz, J., Palomino, J., Riera, M., Del Toro, M.D., Granados, A., Ariza, X., *et al.* (2010). Acute haematogenous prosthetic joint infection: prospective evaluation of medical and surgical management. Clinical microbiology and infection : the official publication of the European Society of Clinical Microbiology and Infectious Diseases *16*, 1789-1795 DOI: 10.1111/j.1469-0691.2010.03157.x.

Rogers, K.L., Fey, P.D., and Rupp, M.E. (2009). Coagulase-negative staphylococcal infections. Infectious disease clinics of North America *23*, 73-98 DOI: 10.1016/j.idc.2008.10.001.

Rossmann, F.S., Racek, T., Wobser, D., Puchalka, J., Rabener, E.M., Reiger, M., Hendrickx, A.P., Diederich, A.K., Jung, K., Klein, C., *et al.* (2015). Phage-mediated dispersal of biofilm and distribution of bacterial virulence genes is induced by quorum sensing. PLoS pathogens *11*, e1004653 DOI: 10.1371/journal.ppat.1004653.

Rozdzinski, E., Marre, R., Susa, M., Wirth, R., and Muscholl-Silberhorn, A. (2001). Aggregation substance-mediated adherence of *Enterococcus faecalis* to immobilized extracellular matrix proteins. Microbial pathogenesis *30*, 211-220 DOI: 10.1006/mpat.2000.0429.

Rui, H., Liu, Y., Lin, G., Hao, C., Ying, Z., and Dian-ming, J. (2016). Management of acute hematogenous infection following total knee arthroplasty: A case series of 11 patients. Orthopaedic Surgery *8*, 475-482 DOI: doi:10.1111/os.12297.

Ryall, B., Davies, J.C., Wilson, R., Shoemark, A., and Williams, H.D. (2008). *Pseudomonas aeruginosa*, cyanide accumulation and lung function in CF and non-CF bronchiectasis patients. The European respiratory journal *32*, 740-747 DOI: 10.1183/09031936.00159607.

Ryder, C., Byrd, M., and Wozniak, D.J. (2007). Role of polysaccharides in *Pseudomonas aeruginosa* biofilm development. Current opinion in microbiology *10*, 644-648 DOI: 10.1016/j.mib.2007.09.010.

Sandoe, J.A., Witherden, I.R., Cove, J.H., Heritage, J., and Wilcox, M.H. (2003). Correlation between enterococcal biofilm formation in vitro and medical-device-related infection potential in vivo. Journal of medical microbiology *52*, 547-550 DOI: 10.1099/jmm.0.05201-0.

Santos, V., and Hirshfield, I. (2016). The physiological and molecular characterization of a Small Colony Variant of *Escherichia coli* and its phenotypic rescue. PloS one *11*, e0157578 DOI: 10.1371/journal.pone.0157578.

Schaller, M., Borelli, C., Korting, H.C., and Hube, B. (2005). Hydrolytic enzymes as virulence factors of *Candida albicans*. Mycoses *48*, 365-377 DOI: 10.1111/j.1439-0507.2005.01165.x.

Schurks, N., Wingender, J., Flemming, H.C., and Mayer, C. (2002). Monomer composition and sequence of alginates from *Pseudomonas aeruginosa*. International journal of biological macromolecules *30*, 105-111.

Schwartz, K., Sekedat, M.D., Syed, A.K., O'Hara, B., Payne, D.E., Lamb, A., and Boles, B.R. (2014). The AgrD N-terminal leader peptide of Staphylococcus aureus has cytolytic and amyloidogenic properties. Infection and immunity *82*, 3837-3844 DOI: 10.1128/IAI.02111-14.

Schwartz, K., Syed, A.K., Stephenson, R.E., Rickard, A.H., and Boles, B.R. (2012). Functional amyloids composed of phenol soluble modulins stabilize *Staphylococcus aureus* biofilms. PLoS pathogens *8*, e1002744 DOI: 10.1371/journal.ppat.1002744.

Selders, G.S., Fetz, A.E., Radic, M.Z., and Bowlin, G.L. (2017). An overview of the role of neutrophils in innate immunity, inflammation and host-biomaterial integration. Regenerative biomaterials *4*, 55-68 DOI: 10.1093/rb/rbw041.

Sendi, P., Banderet, F., Graber, P., and Zimmerli, W. (2011). Periprosthetic joint infection following *Staphylococcus aureus* bacteremia. The Journal of infection *63*, 17-22 DOI: 10.1016/j.jinf.2011.05.005.

Shepherd, N.E., Harrison, R.S., and Fairlie, D.P. (2012). Targeting quorum sensing and competence stimulation for antimicrobial chemotherapy. Current drug targets *13*, 1348-1359.

Sillanpaa, J., Nallapareddy, S.R., Prakash, V.P., Qin, X., Hook, M., Weinstock, G.M., and Murray, B.E. (2008). Identification and phenotypic characterization of a second collagen adhesin, Scm, and genome-based identification and analysis of 13 other predicted MSCRAMMs, including four distinct pilus loci, in *Enterococcus faecium*. Microbiology *154*, 3199-3211 DOI: 10.1099/mic.0.2008/017319-0.

Sillanpaa, J., Xu, Y., Nallapareddy, S.R., Murray, B.E., and Hook, M. (2004). A family of putative MSCRAMMs from *Enterococcus faecalis*. Microbiology *150*, 2069-2078 DOI: 10.1099/mic.0.27074-0.

Silva, S., Negri, M., Henriques, M., Oliveira, R., Williams, D.W., and Azeredo, J. (2012). *Candida glabrata*, *Candida parapsilosis* and *Candida tropicalis*: biology, epidemiology, pathogenicity and antifungal resistance. FEMS Microbiol Rev *36*, 288-305 DOI: 10.1111/j.1574-6976.2011.00278.x.

Singh, K.V., Nallapareddy, S.R., Nannini, E.C., and Murray, B.E. (2005). Fsr-independent production of protease(s) may explain the lack of attenuation of an *Enterococcus faecalis* fsr mutant versus a gelE-sprE mutant in induction of endocarditis. Infection and immunity *73*, 4888-4894 DOI: 10.1128/IAI.73.8.4888-4894.2005.

Sperandio, V., Torres, A.G., Jarvis, B., Nataro, J.P., and Kaper, J.B. (2003). Bacteria-host communication: the language of hormones. Proceedings of the National Academy of Sciences of the United States of America *100*, 8951-8956 DOI: 10.1073/pnas.1537100100.

Speziale, P., Pietrocola, G., Foster, T.J., and Geoghegan, J.A. (2014). Protein-based biofilm matrices in Staphylococci. Frontiers in cellular and infection microbiology *4*, 171 DOI: 10.3389/fcimb.2014.00171.

Srikantha, T., Daniels, K.J., Pujol, C., Kim, E., and Soll, D.R. (2013). Identification of genes upregulated by the transcription factor Bcr1 that are involved in impermeability, impenetrability, and drug resistance of *Candida albicans* a/alpha biofilms. Eukaryotic cell *12*, 875-888 DOI: 10.1128/EC.00071-13.

Stroeher, U.H., Paton, A.W., Ogunniyi, A.D., and Paton, J.C. (2003). Mutation of luxS of *Streptococcus pneumoniae* affects virulence in a mouse model. Infection and immunity *71*, 3206-3212.

Subbiahdoss, G., Kuijer, R., Grijpma, D.W., van der Mei, H.C., and Busscher, H.J. (2009). Microbial biofilm growth vs. tissue integration: "the race for the surface" experimentally studied. Acta biomaterialia *5*, 1399-1404 DOI: 10.1016/j.actbio.2008.12.011.

Sussmuth, S.D., Muscholl-Silberhorn, A., Wirth, R., Susa, M., Marre, R., and Rozdzinski, E. (2000). Aggregation substance promotes adherence, phagocytosis, and intracellular survival of *Enterococcus faecalis* within human macrophages and suppresses respiratory burst. Infection and immunity *68*, 4900-4906.

Szmigielski, S., Prevost, G., Monteil, H., Colin, D.A., and Jeljaszewicz, J. (1999). Leukocidal toxins of staphylococci. Zentralblatt fur Bakteriologie : international journal of medical microbiology *289*, 185-201.

Talpaert, M.J., Balfour, A., Stevens, S., Baker, M., Muhlschlegel, F.A., and Gourlay, C.W. (2015). Candida biofilm formation on voice prostheses. Journal of medical microbiology *64*, 199-208 DOI: 10.1099/jmm.0.078717-0.

Tande, A.J., Palraj, B.R., Osmon, D.R., Berbari, E.F., Baddour, L.M., Lohse, C.M., Steckelberg, J.M., Wilson, W.R., and Sohail, M.R. (2016). Clinical Presentation, Risk Factors, and Outcomes of Hematogenous Prosthetic Joint Infection in Patients with *Staphylococcus aureus* Bacteremia. The American journal of medicine *129*, 221 e211-220 DOI: 10.1016/j.amjmed.2015.09.006.

Tande, A.J., and Patel, R. (2014). Prosthetic joint infection. Clinical microbiology reviews *27*, 302-345 DOI: 10.1128/CMR.00111-13.

Teng, F., Singh, K.V., Bourgogne, A., Zeng, J., and Murray, B.E. (2009). Further characterization of the epa gene cluster and Epa polysaccharides of *Enterococcus faecalis*. Infection and immunity *77*, 3759-3767 DOI: 10.1128/IAI.00149-09.

Teo, A.J.T., Mishra, A., Park, I., Kim, Y.-J., Park, W.-T., and Yoon, Y.-J. (2016). Polymeric Biomaterials for Medical Implants and Devices. ACS Biomaterials Science & Engineering *2*, 454-472 DOI: 10.1021/acsbiomaterials.5b00429.

Tettelin, H., Masignani, V., Cieslewicz, M.J., Eisen, J.A., Peterson, S., Wessels, M.R., Paulsen, I.T., Nelson, K.E., Margarit, I., Read, T.D., *et al.* (2002). Complete genome sequence and comparative genomic analysis of an emerging human pathogen, serotype V *Streptococcus agalactiae*.

Proceedings of the National Academy of Sciences of the United States of America *99*, 12391-12396 DOI: 10.1073/pnas.182380799.

Thurlow, L.R., Hanke, M.L., Fritz, T., Angle, A., Aldrich, A., Williams, S.H., Engebretsen, I.L., Bayles, K.W., Horswill, A.R., and Kielian, T. (2011). *Staphylococcus aureus* biofilms prevent macrophage phagocytosis and attenuate inflammation in vivo. Journal of immunology *186*, 6585-6596 DOI: 10.4049/jimmunol.1002794.

Toledo-Arana, A., Valle, J., Solano, C., Arrizubieta, M.J., Cucarella, C., Lamata, M., Amorena, B., Leiva, J., Penades, J.R., and Lasa, I. (2001). The enterococcal surface protein, Esp, is involved in *Enterococcus faecalis* biofilm formation. Appl Environ Microbiol *67*, 4538-4545.

Toth, I., Herault, F., Beutin, L., and Oswald, E. (2003). Production of cytolethal distending toxins by pathogenic Escherichia coli strains isolated from human and animal sources: establishment of the existence of a new cdt variant (Type IV). Journal of clinical microbiology *41*, 4285-4291.

Tseng, B.S., Reichhardt, C., Merrihew, G.E., Araujo-Hernandez, S.A., Harrison, J.J., MacCoss, M.J., and Parsek, M.R. (2018). A Biofilm Matrix-Associated Protease Inhibitor Protects *Pseudomonas aeruginosa* from Proteolytic Attack. mBio *9* DOI: 10.1128/mBio.00543-18.

Turnbull, L., Toyofuku, M., Hynen, A.L., Kurosawa, M., Pessi, G., Petty, N.K., Osvath, S.R., Carcamo-Oyarce, G., Gloag, E.S., Shimoni, R., *et al.* (2016). Explosive cell lysis as a mechanism for the biogenesis of bacterial membrane vesicles and biofilms. Nature communications *7*, 11220 DOI: 10.1038/ncomms11220.

Van Houdt, R., Aertsen, A., Moons, P., Vanoirbeek, K., and Michiels, C.W. (2006). N-acyl-L-homoserine lactone signal interception by *Escherichia coli*. FEMS microbiology letters *256*, 83-89 DOI: 10.1111/j.1574-6968.2006.00103.x.

Vats, N., and Lee, S.F. (2000). Active detachment of *Streptococcus mutans* cells adhered to epon-hydroxylapatite surfaces coated with salivary proteins in vitro. Archives of oral biology *45*, 305-314.

Veenstra, G.J., Cremers, F.F., van Dijk, H., and Fleer, A. (1996). Ultrastructural organization and regulation of a biomaterial adhesin of *Staphylococcus epidermidis*. Journal of bacteriology *178*, 537-541.

Vergidis, P., and Patel, R. (2012). Novel approaches to the diagnosis, prevention, and treatment of medical device-associated infections. Infectious disease clinics of North America *26*, 173-186 DOI: 10.1016/j.idc.2011.09.012.

Vickery, K., Hu, H., Jacombs, A.S., Bradshaw, D.A., and Deva, A.K. (2013). A review of bacterial biofilms and their role in device-associated infection. Healthcare infection *18*, 61-66 DOI: https://doi.org/10.1071/HI12059.

von Eiff, C., Jansen, B., Kohnen, W., and Becker, K. (2005). Infections associated with medical devices: pathogenesis, management and prophylaxis. Drugs *65*, 179-214.

Vorkapic, D., Pressler, K., and Schild, S. (2016). Multifaceted roles of extracellular DNA in bacterial physiology. Current genetics *62*, 71-79 DOI: 10.1007/s00294-015-0514-x.

Vroman, L. (1988). The life of an artificial device in contact with blood: initial events and their effect on its final state. Bulletin of the New York Academy of Medicine *64*, 352-357.

Vuong, C., Durr, M., Carmody, A.B., Peschel, A., Klebanoff, S.J., and Otto, M. (2004a). Regulated expression of pathogen-associated molecular pattern molecules in *Staphylococcus epidermidis*: quorum-sensing determines pro-inflammatory capacity and production of phenol-soluble modulins. Cellular microbiology *6*, 753-759 DOI: 10.1111/j.1462-5822.2004.00401.x.

Vuong, C., Gerke, C., Somerville, G.A., Fischer, E.R., and Otto, M. (2003). Quorum-sensing control of biofilm factors in *Staphylococcus epidermidis*. The Journal of infectious diseases *188*, 706-718 DOI: 10.1086/377239.

Vuong, C., Kocianova, S., Yao, Y., Carmody, A.B., and Otto, M. (2004b). Increased colonization of indwelling medical devices by quorum-sensing mutants of *Staphylococcus epidermidis* in vivo. The Journal of infectious diseases *190*, 1498-1505 DOI: 10.1086/424487.

Vuong, C., Voyich, J.M., Fischer, E.R., Braughton, K.R., Whitney, A.R., DeLeo, F.R., and Otto, M. (2004c). Polysaccharide intercellular adhesin (PIA) protects *Staphylococcus epidermidis* against major components of the human innate immune system. Cellular microbiology *6*, 269-275.

Walters, M., and Sperandio, V. (2006). Quorum sensing in *Escherichia coli* and *Salmonella*. International journal of medical microbiology : IJMM *296*, 125-131 DOI: 10.1016/j.ijmm.2006.01.041.

Wang, Y., Cheng, L.I., Helfer, D.R., Ashbaugh, A.G., Miller, R.J., Tzomides, A.J., Thompson, J.M., Ortines, R.V., Tsai, A.S., Liu, H., *et al.* (2017). Mouse model of hematogenous implant-related Staphylococcus aureus biofilm infection reveals therapeutic targets. Proceedings of the National Academy of Sciences of the United States of America *114*, E5094-E5102 DOI: 10.1073/pnas.1703427114.

Wenzel, R.P. (1995). Nosocomial candidemia: risk factors and attributable mortality. Clinical infectious diseases : an official publication of the Infectious Diseases Society of America *20*, 1531-1534.

West, S.A., Griffin, A.S., Gardner, A., and Diggle, S.P. (2006). Social evolution theory for microorganisms. Nat Rev Microbiol *4*, 597-607 DOI: nrmicro1461 [pii] 10.1038/nrmicro1461.

Whitchurch, C.B., Tolker-Nielsen, T., Ragas, P.C., and Mattick, J.S. (2002). Extracellular DNA required for bacterial biofilm formation. Science *295*, 1487 DOI: 10.1126/science.295.5559.1487.

Winzer, K., Hardie, K.R., Burgess, N., Doherty, N., Kirke, D., Holden, M.T., Linforth, R., Cornell, K.A., Taylor, A.J., Hill, P.J., *et al.* (2002). LuxS: its role in central metabolism and the in vitro synthesis of 4-hydroxy-5-methyl-3(2H)-furanone. Microbiology *148*, 909-922 DOI: 10.1099/00221287-148-4-909.

Wolz, C., Hohloch, K., Ocaktan, A., Poole, K., Evans, R.W., Rochel, N., Albrecht-Gary, A.M., Abdallah, M.A., and Doring, G. (1994). Iron release from transferrin by pyoverdin and elastase from *Pseudomonas aeruginosa*. Infection and immunity *62*, 4021-4027.

Wood, T.K. (2009). Insights on *Escherichia coli* biofilm formation and inhibition from whole-transcriptome profiling. Environmental microbiology *11*, 1-15 DOI: 10.1111/j.1462-2920.2008.01768.x.

Xavier, K.B., and Bassler, B.L. (2003). LuxS quorum sensing: more than just a numbers game. Current opinion in microbiology *6*, 191-197.

Xu, L., Li, H., Vuong, C., Vadyvaloo, V., Wang, J., Yao, Y., Otto, M., and Gao, Q. (2006). Role of the luxS quorum-sensing system in biofilm formation and virulence of *Staphylococcus epidermidis*. Infection and immunity *74*, 488-496 DOI: 10.1128/IAI.74.1.488-496.2006.

Yanagihara, K., Tomono, K., Kaneko, Y., Miyazaki, Y., Tsukamoto, K., Hirakata, Y., Mukae, H., Kadota, J., Murata, I., and Kohno, S. (2003). Role of elastase in a mouse model of chronic respiratory *Pseudomonas aeruginosa* infection that mimics diffuse panbronchiolitis. Journal of medical microbiology *52*, 531-535 DOI: 10.1099/jmm.0.05154-0.

Yano, J., Yu, A., Fidel, P.L., Jr., and Noverr, M.C. (2016). Transcription Factors Efg1 and Bcr1 Regulate Biofilm Formation and Virulence during Candida albicans-Associated Denture Stomatitis. PloS one *11*, e0159692 DOI: 10.1371/journal.pone.0159692.

Yother, J. (2011). Capsules of *Streptococcus pneumoniae* and other bacteria: paradigms for polysaccharide biosynthesis and regulation. Annu Rev Microbiol *65*, 563-581 DOI: 10.1146/annurev.micro.62.081307.162944.

Zhao, B., van der Mei, H.C., Subbiahdoss, G., de Vries, J., Rustema-Abbing, M., Kuijer, R., Busscher, H.J., and Ren, Y. (2014). Soft tissue integration versus early biofilm formation on different dental implant materials. Dental materials : official publication of the Academy of Dental Materials *30*, 716-727 DOI: 10.1016/j.dental.2014.04.001.

Zimmerli, W., Trampuz, A., and Ochsner, P.E. (2004). Prosthetic-joint infections. N Engl J Med *351*, 1645-1654 DOI: 351/16/1645 [pii] 10.1056/NEJMra040181.

Bacterial Biofilms and Host Immune Response

5

Pradeep Kumar Singh[1], Vivek Kumar Yadav[1], Deepmala Sharma[2], Vishnu Agarwal[1*] and Vandan Nagar[3*]

[1]Department of Biotechnology, Motilal Nehru National Institute of Technology, Allahabad, Prayagraj, India.
[2]Department of Mathematics, National Institute of Technology, Raipur, India.
[3]Food Technology Division, Bhabha Atomic Research Centre, Mumbai, India.

*Corresponding Author: vishnua@mnnit.ac.in, vnagar@barc.gov.in

DOI: https://doi.org/10.21775/9781912530328.05

Abstract

Biofilm includes a group of microbial cells which stick together and remain attached to a surface. Biofilm-associated microbial cells secrete extra-cellular polymeric substances and are embedded in this slimy matrix. Components of the extracellular polymeric substance have a specific role in the development of biofilm. Biofilms are the product of the microbial developmental process which involves different stages, including initial attachment, irreversible attachment, maturation, and dispersion. In many cases, the biofilm mode of microbial growth contributes to antibiotic resistance and escape from the host immune system. Device-associated microbial colonisation plays a vital role in the biofilm-related infections and a large amount of money is being used for the eradication of microbial growth from these devices. Biofilm interacts with the different components of the host immune system and may contribute to the stimulation or suppression of immune response. Different immune cells, including neutrophils, macrophages, T Cells, and B-cells, generate a specific response against biofilms.

Introduction

In nature, microbes exist in both planktonic and biofilm forms. Biofilms have been considered as a complex community of viable and non-viable microorganisms enclosed within an extracellular polymeric substances (EPS), which adhere to the biotic or abiotic substratum and consist of single or multiple microbial species (Vert et al., 2012). The surface association is an efficient means of establishment of biofilm in a favourable microenvironment rather than being swept away by the current. Depending on the availability of substratum, biofilms may either remain attached to the surface or exist in the mobile phase. In multilayer biofilms, the cells of only a single layer are directly attached with the substratum. Genetic studies have revealed that biofilm formation is a multi-step process which requires inter-cellular signalling (Davies et al., 1998; Watnick and Kolter, 1999). Presence of EPS and inter-cellular interactions result in a distinct lifestyle of biofilm as compared to free-living bacterial cells (Vert et al., 2012). In the process of microbial biofilm formation, biofilm associated cells show a distinct gene transcription profile compared to the planktonic cells. (Prigent-Combaret et al., 1999).

Most of the higher organisms, including humans, are colonised by biofilm-forming microorganisms (De Vos, 2015) that may be associated with persistent infections (Costerton et al., 1987) and contamination of medical implants (Shirtliff and Leid, 2009). Several medical devices are susceptible to colonisation and contamination by bacterial cells (Arciola, 2018; Castelli et al., 2006). Several microbial infections have been linked with different medical implant devices, including prosthetic heart valves, orthopaedic implants, intra-vascular catheters, artificial hearts, left ventricular assist devices, cardiac pacemakers, urinary catheters, ocular prostheses, contact lenses and intra-uterine contraceptive devices (Arciola et al., 2018). Biofilm-related infections are hard to treat and they usually manifest as chronic or recurrent infections. Biofilm-associated infections may include various clinical challenges involving chronic inflammation, impaired wound healing, acquired antibiotic resistance, and spread of the infectious emboli (Castelli et al., 2006; Flemming and Wingender, 2010; Hoiby, 2014). The summary of most common bacterial biofilm-associated infections in humans is listed in Table 1.

Hydrated EPS matrix constitutes the major part (~ 80-90%) of the biofilm biomass (Flemming and Wingender, 2010). EPS contributes to the development of different characteristics of biofilm community such as the formation of physical and social interactions, an enhanced rate of gene exchange between cells and increased tolerance to antimicrobials, phagocytes, and host immune mechanisms (Flemming and Wingender, 2010). Water is a major component of the EPS matrix which consist of structural and functional elements of the matrix, including soluble components like polysaccharides, proteins and e-DNA (Flemming and Wingender, 2010) along with insoluble components including amyloids, fimbriae, flagella, pili (Hobley et al., 2015) and cellulose (Serra et al., 2013). EPS associated molecules contribute to the formation of biofilm structure through a continuous and dynamic process, resulting in a spatial organisation in which biofilm cells aggregate in microcolonies (Neu and Lawrence, 2015). EPS matrix of biofilm is mainly composed of polysaccharides, proteins, lipids, and extracellular DNA (e-DNA). Biofilm-associated e-DNA has a significant role in the formation and persistence of the biofilm structure. Cationic polysaccharide such as Pel crosslinks e-DNA in the *Pseudomonas aeruginosa* biofilm matrix (Jennings et al., 2015). In a biofilm system, cellular decay and lysis result in the accumulation of cellular debris, which is further utilised as nutrients by the surviving cells. In a *Bacillus subtilis* biofilm, the e-DNA released from lysed cells are a source of carbon, phosphorus and energy (Pinchuk et al., 2008). *P. aeruginosa* produces extracellular DNases in the biofilm to utilise DNA from the lysed or decaying cells as a nutrient source (Mulcahy et al., 2010).

A recent study has shown that the voids formed in the matrix between micro-colonies facilitate the transport of liquid in the biofilms through pores and channels (Wilking et al., 2013). Biofilm matrix associated structural components have other functions too, for example, a curly protein of *Escherichia coli* biofilm matrix together with cellulose contributes to desiccation tolerance and *Bacillus subtilis* uses hydrophobic protein to form a highly hydrophobic biofilm (Hobley et al., 2015). Other biofilm matrix associated components, including proteinaceous filaments and nanowires, participate in electron transport (Lovley and Malvankar, 2015).

Table 1. List of common bacterial biofilm-associated infections

Biofilm infection or disease	**Bacterial species**
Dental caries	Acidogenic gram-positive cocci (e.g. *Streptococcus*)
Cystic fibrosis pneumonia	*P. aeruginosa* and *Burkholderia cepacia*
Periodontitis	Gram-negative anaerobic oral bacteria
Musculoskeletal infections	Gram-positive cocci (e.g., *Staphylococci*)
Necrotizing fasciitis	Group A *Streptococci*
Osteomyelitis	Various bacterial and fungal species often mixed culture
Melioidosis	*Pseudomonas pseudomallei*
Sutures	*Staphylococcus epidermidis* and *Staphylococcus aureus*
Arteriovenous shunts	*S. epidermidis* and *S. aureus*
Schleral buckles	Gram-positive cocci
Contact lens	*P. aeruginosa* and Gram-positive cocci
Vascular grafts	Gram-positive cocci
Orthopaedic devices	*E. coli,* hemolytic *Streptococci*, *Enterococci*, *Proteus mirabilis*, *P. aeruginosa*
Urinary catheter cystitis	*S. epidermidis, Klebsiella pneumoniae, Enterococcus faecalis, P. mirabilis, P. aeruginosa*
Peritoneal dialysis (CAPD) peritonitis	*S. epidermidis*, *S. aureus*, *Corynebacterium* sp., *Micrococcus* sp., *Enterococcus* sp., *Candida albicans*, Group B *Streptococci*
Hickman catheters	*S. epidermidis* and *C. albicans*

Antibiotic resistance in bacterial biofilm

Resistant bacteria can survive even in the presence of molecules having bactericidal or bacteriostatic property, while the equivalent amount of these compounds can kills or inhibit the growth of susceptible microorganisms. Resistance is generally measured in the form of minimum inhibitory concentration (MIC), which is the lowest concentration of the antimicrobial drug that can inhibit the growth of sensitive bacteria. The emergence of microbial resistance to antimicrobial agents is considered to be due to the mutation or acquisition of antibiotic resistance genes. Antibiotic resistance can be intrinsic so it depends on the wild type genetic material and the inborn properties of the cell (Blair et al., 2011; Cox and Wright, 2013). According to Lewis 2008, antibiotic resistance is attributed to the prevention of antimicrobial agents' interaction with their intended target (Lewis, 2008). Furthermore, antibiotic resistance in biofilm may develop through various mechanisms such as activity of antibiotic efflux pumps, presence of antibiotic sequestering molecules, and matrix associated β-lactamases (Hall and Mah, 2017). In contrast, tolerance is the ability of the microorganism to survive with no observable growth or death in the presence of antimicrobial agents (Hall and Mah, 2017). In case of bacterial biofilm, pathogens utilises both antibiotic resistance and tolerance mechanisms to survive in the presence of antimicrobial molecules (Hall and Mah, 2017). Low penetration of antimicrobial compounds in the biofilm matrix is the most accepted explanation for the biofilm tolerance and resistance properties (Hall and Mah, 2017). In an experimental setting, antibiotic diffusion in a biofilm system depends on various factors, including bacterial strains or species, type of antibiotic, and growth conditions of the biofilm (Hall and Mah, 2017). Recently, it has been proposed that slow diffusion of antimicrobial agents allow bacterial cells to develop resistance phenotype that could contribute to antibiotic tolerance (Tseng et al., 2013). Exopolysaccharides have also been reported to play an important role in the development of antimicrobial resistance in bacterial biofilms, such as *P. aeruginosa* associated exopolysaccharide Psl is essential for the formation of biofilms and it contributes to antibiotic resistance in the initial phase of biofilm formation. (Byrd et al., 2009; Billings et al., 2013). Pel is a glycocalyx structure of *Pseudomonas* biofilm that protects it against gentamicin and tobramycin (Colvin et al., 2011; Franklin et al., 2011). Biofilm matrix enzymes like β-lactamases can degrade antimicrobial and thus prevent their access to respective targets (Anderl et al., 2003). Biofilm matrix e-DNA, both endogenous and exogenous of origin, increases antimicrobial resistance (Chiang et al., 2013). Filamentous phage particles cause death of the bacterial cell leading to release of extracellular DNA and the growth of antimicrobial tolerant small colony variants inside the bacterial biofilms (Webb et al., 2003; Webb et al., 2004). Antibiotic tolerance also develops as a part of the increase in physiological heterogeneity, which occurs in the cells located in the different geographic areas of biofilm (Hall and Mah, 2017). Cell-wall modifying enzymes could also contribute to antimicrobial tolerance. The *dlt ABCD* operon and its homologs mediate D-alanylation of teichoic acid molecules in the cell wall of *Streptococcus mutans* contribute to the development of antimicrobial tolerance (Nilsson et al., 2016). Many other possible mechanisms by which biofilm develops tolerance against antimicrobials are glycosyltransferase, transcriptional regulators, efflux pumps, quorum sensing, genetic diversity, persistent cells and multispecies interactions (Hall and Mah, 2017). Most common mechanisms by which biofilm may attain antibiotic resistance is illustrated in Figure 1.

Interactions between biofilm and immune cells

Bacterial cells make biofilms to prevent recognition from the host immune system. Some previous studies focused mainly on the interaction between biofilm mode of bacterial cells and the immune system. The immune response against biofilms is very complicated and ambiguous. Biofilms may suppress or stimulate the host immune system. These responses are likely to depend on the various factors such as bacterial species, location of the biofilm in the host, and status of the individual's immune system. The host immune system interacts directly or indirectly with bacterial cells through their secretory components. The formation of the biofilm matrix involves multiple bacterial cell-secreted products, while others may have different functions. All of these secretory components can be immunogenic. Host immune system, which involves a complicated interplay of cellular signalling and immune cells, has been divided into two main categories: innate and adaptive immune systems. The innate immune system consists of the epithelial layer, complement systems, host microbiome and a number of different cells including both phagocytic and non-phagocytic cells.

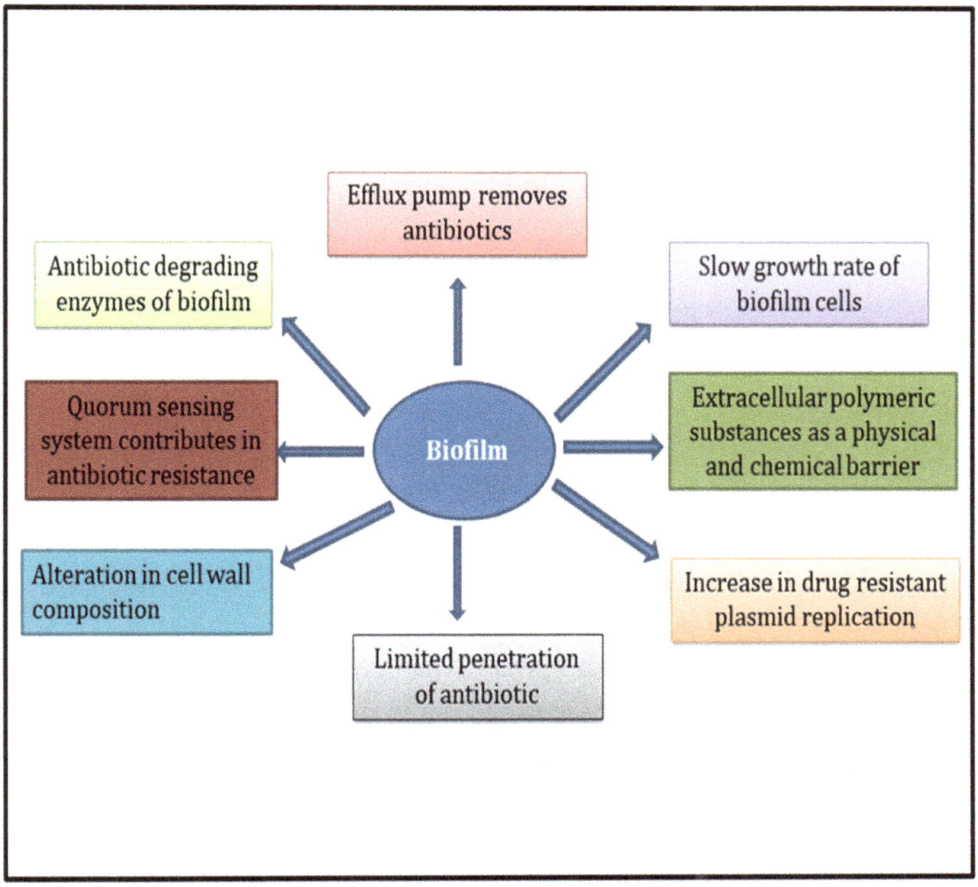

Figure 1. Mechanisms determining bacterial biofilm resistance to antimicrobial agents.

Cells involved in the innate immune system mainly utilise pattern recognition receptors (PRRs) to recognise conserved molecular pattern specific to particular pathogen, for example peptidoglycan and lipopolysaccharides (LPS). These immune cells play a crucial role in the prevention of bacterial invaders, and induce the secretion of several inflammatory mediators including cytokines and chemokines. The following segment emphases on the interaction of primary innate immune cells such as neutrophils and macrophages with bacterial biofilms.

Biofilm interaction with neutrophil

Neutrophils are the first cells that respond to the inflammatory reaction and reach to the site of infection (Summers et al., 2010). Tissue-specific cells like epithelial cells, mast cells, and macrophages attracts more neutrophils to the site of infection by releasing various chemo-attractants such as CXCL1/CXCL2 (De Filippo et al., 2013), IL-8 and activating cytokine IL-6. Polymorphonuclear neutrophils (PMNs) reach to the infected site through the chemotaxis movement and start the process of phagocytosis and release pro-inflammatory cytokines such as TNF-α, IL-1β and IL-8 (Witko-Sarsat et al., 2000). These pro-inflammatory cytokines help to attract more neutrophils at the site of infection and generate inflammation. PMN inactivate bacterial cells through the various mechanisms, such as the secretion of antimicrobial granules (lysozyme, lactoferrin and defensins), production of ROS, phagocytosis and release of neutrophil extracellular traps (NETs) (Amulic et al., 2012). The success of phagocytosis depends on the opsonisation of the bacterial cells. Opsonisation is the process by which microbes get coated with an opsonin that helps in the removal of the bacterial cell via the activation of the complement system or phagocytosis (Van Kesse et al., 2014). The bacterial cells opsonised with IgG and C3b, get recognised by Fc-gamma receptor and complement receptor, respectively, which may result in PMN mediated phagocytosis. After phagocytosis, PMN apoptosis resolves inflammation and macrophages further remove these apoptotic cells. (Cox et al., 1995).

In most of the cases, opsonisation is essential for the clearance of bacteria by phagocytosis, very less deposition of IgG on biofilm was observed on co-incubation with IgG resulting in delayed phagocytosis by PMNs (Günther et al., 2009). Most of the PMNs and biofilms interaction studies have been conducted with *P. aeruginosa* or *Staphylococcus sp*. It has been demonstrated that the biofilm of *S. aureus*, grown for different days (2, 6 and 15 days), was reduced by PMN mediated phagocytosis (Günther et al., 2009). The clearance of *S. aureus* by PMN was shown to be IL1-β dependent in mice (Günther et al., 2009). In addition, *S. epidermidis* biofilms are resistant to the phagocytosis and killing by PMNs relative to their EPS mutant strain and planktonic cells (Otto, 2009).

The oxidative burst of PMNs was reduced in the presence of *P. aeruginosa* biofilm compared to planktonic cells. (Jensen et al., 1990). In a different study, it was observed that the rate of PMN mediated phagocytosis of the planktonic cells was higher than the biofilm cells (Jesaitis et al., 2003; Alhede et al., 2011). It has been proposed that the inhibition in phagocytosis of biofilm cells by PMNs was due to the presence of rhamnolipid in the biofilm. Rhamnolipids are glycolipids produced by *P. aeruginosa,* which serves as bio-surfactant, and their production is regulated by the quorum sensing (QS) process (Dusane et al., 2010). However, rhamnolipids of both planktonic and biofilm culture, reduce the chemotaxis movement and rapidly kill the neutrophil (Jensen et al., 2007). Furthermore, it was noted that, in pulmonary infection model the inhibitor of rhamnolipid significantly prevent the *P. aeruginosa* biofilm (Jakobsen et al., 2012). The protection of biofilm by rhamnolipid from PMN-mediated phagocytosis is called rhamnolipid shielding.

P. aeruginosa rhlI/rhlR QS pathway is directly associated with the production of rhamnolipid. It is reported that the supernatant of *lasR/rhlR* mutant *P. aeruginosa* biofilm was unable to kill PMNs. Moreover, supernatant from wild-type PAO1 causes lysis of whole blood; whereas, *lasR/rhlR* mutants were unable to do so. Sputum from *P. aeruginosa* infected cystic fibrosis patients causes lysis of PMNs; whereas, no effect was observed in the sputum of uninfected individual (Alhede et al., 2009). The previous report shows that the biofilm of QS mutant strain could be phagocytised efficiently compared to the wild type strain (Bjarnsholt et al., 2005). In contrast, the production of QS-regulated virulence factors is positively regulated in the presence of PMNs (Van Gennip et al., 2012). There are various reports which suggest that PMNs augment *P. aeruginosa* biofilm and EPS after 72 h (Walker et al., 2005). It has also been shown that *P. aeruginosa* used actin and DNA of lysed PMNs and enhanced the thickness of the biofilm (Walker et al., 2005).

Biofilm interaction with macrophages

Macrophages phagocytose the microbes and stimulate the immune system by recruiting PMNs. Monocytes (immature macrophages) are recruited to the site of infection after the phagocytosis of bacteria by PMNs and their subsequent apoptosis. (Dale et al., 2008). Monocytes are recruited by chemokines such as CCL2, CCL7, N-formyl-methionyl peptides and pro-inflammatory cytokines at the site of infection (Shi and Pamer, 2011). At the site of infection, macrophages are engaged in the various immune defence activities, for example, enhanced phagocytosis, antigen presentation and release of various cytokines (Mosser and Edwards, 2008). Macrophages are capable of recognising and helping to absorb foreign particles, including bacterial cells via their surface receptors. Once macrophage engulfs the bacteria, it produces ROS and NO to kill them in the cytoplasm.

Though there are only a few reports available describing the interaction of macrophages with biofilm, a study focused on *S. epidermidis* biofilm, and macrophage has shown that EPS of *S. epidermidis* biofilm, significantly reduces the oxidative burst of rabbit alveolar macrophages (Schommer et al., 2011). Other components of the *S. epidermidis* biofilm, including polysaccharide intercellular adhesion (PIA), Aap, and Embp intracellular adhesins are involved in the immune escape or protection from phagocytosis by macrophages (Schommer et al., 2011). *S. epidermidis* biofilm significantly inhibits NF-κB activation and cause a subsequent reduction in IL-1β release (Cerca et al., 2011).

The alginate of *P. aeruginosa* biofilms prevents macrophages from killing of *P. aeruginosa* cells (Limoli et al., 2015). The production of alginate is increased by two-fold when *P. aeruginosa* biofilm was grown in the presence of macrophage supernatant. *P. aeruginosa* biofilm formation enhances the level of cytokines production in the host, as compared to planktonic cells (Mittal et al., 2006), which may further recruit more macrophages. Besides, macrophage supernatant enhanced the significant viable number of both biofilm and planktonic cells of *P. aeruginosa*. In this regard, Mittal et al. suggested that macrophages can enhance the *P. aeruginosa* biofilm (Mittal et al., 2006). A similar effect was observed when biofilm of *Candida* was enhanced on co-incubation with monocyte or macrophage supernatant (Chandra et al., 2007).

It was reported that *S. aureus* biofilm suppresses the phagocytic activity of macrophages and increases the killing of host cells (Scherr et al., 2015). It depends on *S. aureus* associated alpha toxin and lekocidin AB, because when the bacterial load was decreased in case of *S. aureus* double mutant *lukAB/hlaS*, the number of macrophages increased significantly (Scherr et al., 2015). *S. aureus* stimulates the MyD88-signalling pathway (Hanke et al., 2012). In the mice model of catheter-related infection, bacterial count, fibrosis condition and number of activated macrophages were increased in MyD88 deficient mice; whereas pro-inflammatory cytokines production was reduced (Hanke et al., 2012). Overall, it was concluded that EPS of biofilms significantly challenged the phagocytic activity and ROS generation of macrophages. Importantly, macrophage supernatant may induce the EPS production, and subsequently enhance biofilm formation.

The overall mechanisms by which bacterial biofilm modulates the function of immune cells (neutrophils and macrophages) and escape from host immunity is summarised in Figure 2.

Adaptive immune response against biofilms

The adaptive immune system is composed of both humoral as well as the cell-mediated immune response. They work cooperatively with the innate immune system and induce a memory response to clear the infection. Humoral immunity is mediated by the B cell, which produces the antibody to neutralise the bacterial toxins and opsonise the bacteria to facilitate the phagocytic process (Stephen et al., 2010). Cell-Mediated immunity is mainly regulated by T cells. CD4+ T cells are T-helper cells that facilitate the Th1, Th2, and Th17 reactions, while CD8+ T cells are called cytotoxic T cells. T cells recognise the antigen presented by antigen presenting cells like macrophages and dendritic cells, and further, get activated and converted to the mature T cells. The cell-mediated immunity is Th1 mediated and gets activated against intracellular pathogens; while humoral immunity is mainly Th2 dependent, which mainly targets the extracellular pathogens (Heim et al., 2015). T helper cell is important for clearing extracellular bacteria, and recently it has been proven that they participate in biofilm-associated inflammations. The following section will discuss the interaction of biofilm with an adaptive immune response. A high level of antibody and IL-6 production was seen in the *S. aureus* infection model of porcine osteomyelitis (Jensen et al., 2015). As IL-6 is associated with both Th1 and Th17 reactions, it indicates that the biofilm of *S. aureus* can escape Th2 related killing (Jensen et al., 2015).

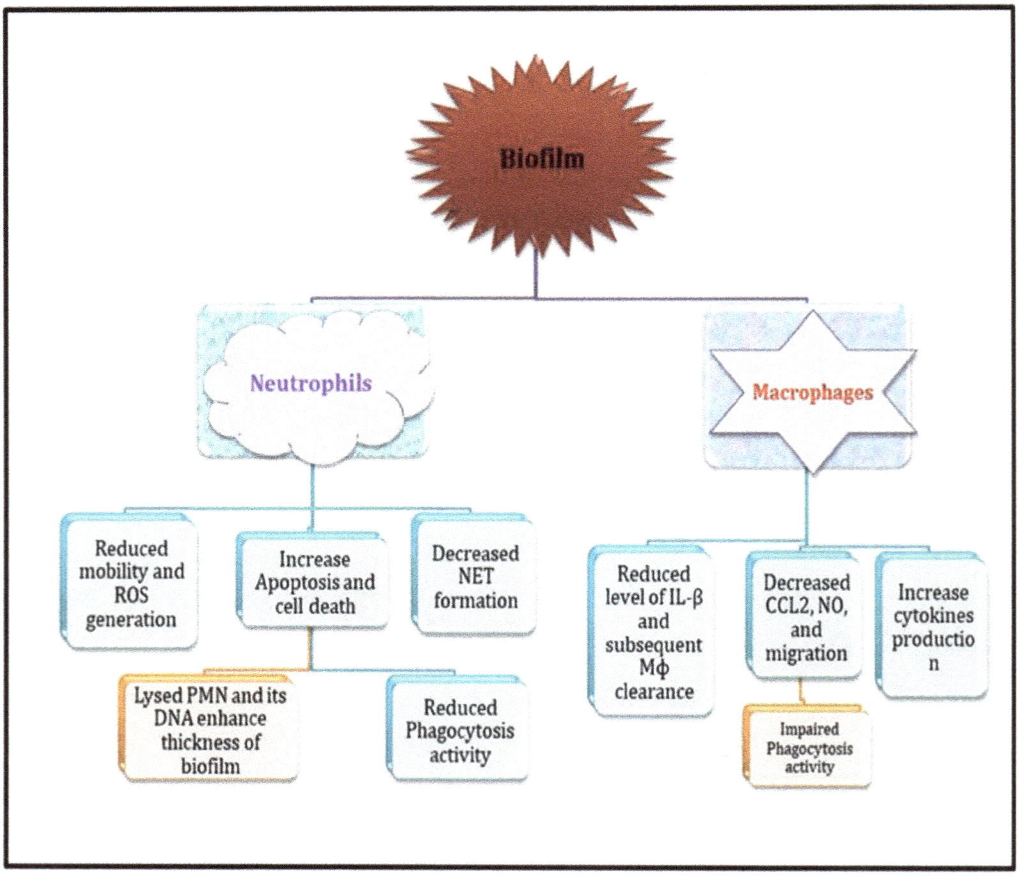

Figure 2. Escape mechanism of bacterial biofilm from neutrophils and macrophages .

In a murine implant model, the antibody response against *S. epidermidis* biofilm was studied, and it was noted that the level of Th1 associated antibody isotype IgG2b was high during early infection (up to 14 days), as the infection remained up to 28 days, the level of Th2 associated antibody type IgG1 was increased (Prabhakara et al., 2011). It has been established that during bacterial infection, myeloid-derived suppressor cells (MDSCs) inhibit T cell function leading to chronic infection in the host. The involvement of MDSCs in *S. aureus* biofilm infection was studied and it was found that the levels of IL-10 and Arg-1 (MDSCs activator) were augmented and the numbers of T cells were decreased (Heim et al., 2014). The immune-depletion of MDSCs enhanced the activity of macrophage against *S. aureus* biofilms clearance. When MDSCs was exogenously added to p40 knockout mice, the bacterial load was increased during infection (Heim et al., 2015). These reports highlighted that in what way *S. aureus* biofilm can alter the host immunity to escape from the host immune system. A large number of studies on interaction of *S. aureus* biofilms with the adaptive immune system have been shown the importance of cell-mediated and humoral immunity. However, further studies are needed to decipher the involvement of the adaptive immune response against various single and multi-species biofilms.

Conclusion.

Treatment of biofilm-associated infections is currently a complicated and challenging task for clinicians and microbiologists. Antibiotics alone are often insufficient to overcome biofilm-related infections. However, the progress of research in the field of biofilms has provided more comprehensive knowledge for better understanding of the nature of microbial biofilms, which has helped and will continue to support efforts for fighting biofilm infections. Currently, treatment of biofilm infections needs a significant collaboration in clinical microbiology, internal medicine, surgery, pharmacology and basic sciences. It is believed that the treatment of biofilm at present should comprise of exclusion of infected indwelling medical devices, selection of sensitive antibiotics and early administration of antibiotics in combination with anti-QS treatment or biofilm dispersing agents.

Web Resources

https://www.nature.com/subjects/biofilms

https://www.cdc.gov/index.htm

http://www.biofilm.montana.edu/biofilm-basics/index.html

https://biofilm.ku.dk

References

Alhede, M., Bjarnsholt, T., Jensen, P.Ø., Phipps, R.K., Moser, C., Christophersen, L., Christensen, L.D., van Gennip, M., Parsek, M., and Høiby, N. (2009). *Pseudomonas aeruginosa* recognises and responds aggressively to the presence of polymorphonuclear leukocytes. Microbiol *155*, 3500–3508 DOI: 10.1099/mic.0.031443-0.

Alhede, M., Kragh, K.N., Qvortrup, K., Allesen-Holm, M., van Gennip, M., Christensen, L.D., Jensen, P.Ø., Nielsen, A.K., Parsek, M., and Wozniak, D. (2011). Phenotypes of non-attached *Pseudomonas aeruginosa* aggregates resemble surface-attached biofilm. PLoS One *6*, e27943 DOI: 10.1371/journal.pone.0027943.

Amulic, B., Cazalet, C., Hayes, G.L., Metzler, K.D., and Zychlinsky, A. (2012). Neutrophil function: from mechanisms to disease. Annu Rev Immunol *30*, 459–489 DOI: 10.1146/annurev-immunol-020711-074942.

Anderl, J.N., Zahller, J., Roe, F., and Stewart, P.S. (2003). Role of nutrient limitation and stationary-phase existence in *Klebsiella pneumoniae* biofilm resistance to ampicillin and ciprofloxacin. Antimicrob Agents Chemother *47*, 1251–1256 DOI: 10.1128/aac.47.4.1251-1256.2003.

Arciola, C.R., Campoccia, D., and Montanaro, L. (2018). Implant infections: adhesion, biofilm formation and immune evasion. Nat Rev Microbiol 1–13 DOI: 10.1038/s41579-018-0019-y.

Billings, N., Millan, M.R., Caldara, M., Rusconi, R., Tarasova, Y., Stocker, R., and Ribbeck, K. (2013). The extracellular matrix component Psl provides fast-acting antibiotic defense in *Pseudomonas aeruginosa* biofilms. PLoS Pathog *9*, e1003526 DOI: 10.1371/journal.ppat.1003526.

Bjarnsholt, T., Jensen, P.Ø., Burmølle, M., Hentzer, M., Haagensen, J.A.J., Hougen, H.P., Calum, H., Madsen, K.G., Moser, C., and Molin, S. (2005).

Pseudomonas aeruginosa tolerance to tobramycin, hydrogen peroxide and polymorphonuclear leukocytes is quorum-sensing dependent. Microbiol *151*, 373–383 DOI: 10.1099/mic.0.27463-0.

Blair, J.M.A., Webber, M.A., Baylay, A.J., Ogbolu, D.O., and Piddock, L.J.V. (2011). Molecular mechanisms of antibiotic resistance. Nat Rev Microbiol *47*, 4055–4061 DOI: 10.1038/nrmicro3380.

Byrd, M.S., Sadovskaya, I., Vinogradov, E., Lu, H., Sprinkle, A.B., Richardson, S.H., Ma, L., Ralston, B., Parsek, M.R., Anderson, E.M., et al. (2009). Genetic and biochemical analyses of the *Pseudomonas aeruginosa* Psl exopolysaccharide reveal overlapping roles for polysaccharide synthesis enzymes in Psl and LPS production. Mol Microbiol *73*, 622–638 DOI: 10.1111/j.1365-2958.2009.06795.x.

Castelli, P., Caronno, R., Ferrarese, S., Mantovani, V., Piffaretti, G., Tozzi, M., Lomazzi, C., Rivolta, N., and Sala, A. (2006). New trends in prosthesis infection in cardiovascular surgery. Surg Infect *7 Suppl 2*, S45-7 DOI: 10.1089/sur.2006.7.s2-45.

Cerca, F., Andrade, F., França, Â., Andrade, E.B., Ribeiro, A., Almeida, A.A., Cerca, N., Pier, G., Azeredo, J., and Vilanova, M. (2011). *Staphylococcus epidermidis* biofilms with higher proportions of dormant bacteria induce a lower activation of murine macrophages. J Med Microbiol *60*, 1717–1724 DOI: 10.1099/jmm.0.031922-0.

Chandra, J., McCormick, T.S., Imamura, Y., Mukherjee, P.K., and Ghannoum, M.A. (2007). Interaction of *Candida albicans* with adherent human peripheral blood mononuclear cells increases *C. albicans* biofilm formation and results in differential expression of pro-and anti-inflammatory cytokines. Infect Immun *75*, 2612–2620 DOI: 10.1128/IAI.01841-06.

Chiang, W.-C., Nilsson, M., Jensen, P.Ø., Høiby, N., Nielsen, T.E., Givskov, M., and Tolker-Nielsen, T. (2013). Extracellular DNA shields against aminoglycosides in *Pseudomonas aeruginosa* biofilms. Antimicrob Agents Chemother *57*, 2352–2361 DOI: 10.1128/AAC.00001-13.

Colvin, K.M., Gordon, V.D., Murakami, K., Borlee, B.R., Wozniak, D.J., Wong, G.C.L., and Parsek, M.R. (2011). The pel polysaccharide can serve a structural and protective role in the biofilm matrix of *Pseudomonas aeruginosa*. PLoS Pathog *7*, e1001264 DOI: 10.1371/journal.ppat.1001264.

Costerton, J.W., Cheng, K.J., Geesey, G.G., Ladd, T.I., Nickel, J.C., Dasgupta, M., and Marrie, T.J. (1987). Bacterial biofilms in nature and disease. Annu Rev Microbiol *41*, 435–464 DOI: 10.1146/annurev.mi.41.100187.002251.

Cox, G., and Wright, G.D. (2013). Intrinsic antibiotic resistance: Mechanisms, origins, challenges and solutions. Int J Med Microbiol *303*, 287–292 DOI: 10.1016/j.ijmm.2013.02.009.

Cox, G., Crossley, J., and Xing, Z. (1995). Macrophage engulfment of apoptotic neutrophils contributes to the resolution of acute pulmonary inflammation *in vivo*. Am J Respir Cell Mol Biol *12*, 232–237 DOI: 10.1165/ajrcmb.12.2.7865221.

Dale, D.C., Boxer, L., and Liles, W.C. (2008). The phagocytes: neutrophils and monocytes. Blood *112*, 935–945 DOI: 10.1182/blood-2007-12-077917.

Davies, D.G., Parsek, M.R., Pearson, J.P., Iglewski, B.H., Costerton, J.W., and Greenberg, E.P. (1998). The involvement of cell-to-cell signals in the development of a bacterial biofilm. Science *280 (5361)*, 295–298.

Dusane, D.H., Zinjarde, S.S., Venugopalan, V.P., Mclean, R.J.C., Weber, M.M., and Rahman, P.K.S.M. (2010). Quorum sensing: implications on rhamnolipid biosurfactant production. Biotechnol Genet Eng Rev *27*, 159–184.

De Filippo, K., Dudeck, A., Hasenberg, M., Nye, E., van Rooijen, N., Hartmann, K., Gunzer, M., Roers, A., and Hogg, N. (2013). Mast cell and macrophage chemokines CXCL1/CXCL2 control the early stage of neutrophil recruitment during tissue inflammation. Blood *121*, 4930–4937 DOI: 10.1182/blood-2013-02-486217.

Flemming, H.C., and Wingender, J. (2010). The biofilm matrix. Nat Rev Microbiol *8*, 623–633 DOI: 10.1038/nrmicro2415.

Franklin, M.J., Nivens, D.E., Weadge, J.T., and Howell, P.L. (2011). Biosynthesis of the *Pseudomonas aeruginosa* extracellular polysaccharides, alginate, Pel, and Psl. Front Microbiol 2, 167 DOI: 10.3389/fmicb.2011.00167.

Van Gennip, M., Christensen, L.D., Alhede, M., Qvortrup, K., Jensen, P.Ø., Høiby, N., Givskov, M., and Bjarnsholt, T. (2012). Interactions between polymorphonuclear leukocytes and *Pseudomonas aeruginosa* biofilms on silicone implants *in vivo*. Infect Immun *80*, 2601–2607 DOI: 10.1128/IAI.06215-11.

Günther, F., Wabnitz, G.H., Stroh, P., Prior, B., Obst, U., Samstag, Y., Wagner,

C., and Hänsch, G.M. (2009). Host defence against *Staphylococcus aureus* biofilms infection: Phagocytosis of biofilms by polymorphonuclear neutrophils (PMN). Mol Immunol *46*, 1805–1813 DOI: 10.1016/j.molimm.2009.01.020.

Hall, C.W., and Mah, T.-F. (2017). Molecular mechanisms of biofilm-based antibiotic resistance and tolerance in pathogenic bacteria. FEMS Microbiol Rev *41*, 276-301 DOI: 10.1093/femsre/fux010.

Hanke, M.L., Angle, A., and Kielian, T. (2012). MyD88-dependent signaling influences fibrosis and alternative macrophage activation during *Staphylococcus aureus* biofilm infection. PLoS One *7*, e42476 DOI: 10.1371/journal.pone.0042476.

Heim, C.E., Vidlak, D., Scherr, T.D., Kozel, J.A., Holzapfel, M., Muirhead, D.E., and Kielian, T. (2014). Myeloid-derived suppressor cells contribute to *Staphylococcus aureus* orthopedic biofilm infection. J Immunol *192*, 3778–3792

DOI: 10.4049/jimmunol.1303408.

Heim, C.E., Vidlak, D., Scherr, T.D., Hartman, C.W., Garvin, K.L., and Kielian, T. (2015). IL-12 promotes myeloid-derived suppressor cell recruitment and bacterial persistence during *Staphylococcus aureus* orthopedic implant infection. J Immunol *194*, 3861–3872 DOI: 10.4049/jimmunol.1402689.

Hobley, L., Harkins, C., MacPhee, C.E., and Stanley-Wall, N.R. (2015). Giving structure to the biofilm matrix: An overview of individual strategies and emerging common themes. FEMS Microbiol Rev *39*, 649–669 DOI: 10.1093/femsre/fuv015.

Hoiby, N., Bjarnsholt, T., Moser, C., Bassi, G. L., Coenye, T., & Donelli, G. (2015). ESCMID guideline for the diagnosis and treatment of biofilm infections 2014. Clin Microbiol Infect *21,* Suppl 1:S1-S25. DOI: 10.1016/j.cmi.2014.10.024.

Jakobsen, T.H., Van Gennip, M., Phipps, R.K., Shanmugham, M.S., Christensen, L.D., Alhede, M., Skindersoe, M.E., Rasmussen, T.B., Friedrich, K., and Uthe, F. (2012). Ajoene, a sulfur-rich molecule from garlic, inhibits genes controlled by quorum sensing. Antimicrob Agents Chemother *56*, 2314–2325 DOI: 10.1128/AAC.05919-11.

Jennings, L.K., Storek, K.M., Ledvina, H.E., Coulon, C., Marmont, L.S., Sadovskaya, I., Secor, P.R., Tseng, B.S., Scian, M., Filloux, A., *et al.* (2015).

Pel is a cationic exopolysaccharide that cross-links extracellular DNA in the *Pseudomonas aeruginosa* biofilm matrix. Proc Natl Acad Sci *112*, 11353–11358 DOI: 10.1073/PNAS.1503058112.

Jensen, E.T., Kharazmi, A., Lam, K., Costerton, J.W., and Høiby, N. (1990). Human polymorphonuclear leukocyte response to *Pseudomonas aeruginosa* grown in biofilms. Infect Immun *58*, 2383–2385.

Jensen, L.K., Jensen, H.E., Koch, J., Bjarnsholt, T., Eickhardt, S., and Shirtliff, M. (2015). Specific antibodies to *Staphylococcus aureus* biofilm are present in serum from pigs with osteomyelitis. In Vivo *29*, 555–560.

Jensen, P.Ø., Bjarnsholt, T., Phipps, R., Rasmussen, T.B., Calum, H., Christoffersen, L., Moser, C., Williams, P., Pressler, T., and Givskov, M. (2007). Rapid necrotic killing of polymorphonuclear leukocytes is caused by quorum-sensing-controlled production of rhamnolipid by *Pseudomonas aeruginosa*. Microbiol *153*, 1329–1338 DOI: 10.1099/mic.0.2006/003863-0.

Jesaitis, A.J., Franklin, M.J., Berglund, D., Sasaki, M., Lord, C.I., Bleazard, J.B., Duffy, J.E., Beyenal, H., and Lewandowski, Z. (2003). Compromised host defense on *Pseudomonas aeruginosa* biofilms: characterization of neutrophil and biofilm interactions. J Immunol *171*, 4329–4339 DOI: 10.4049/jimmunol.171.8.4329.

Van Kesse, K.P.M., Bestebroer, J., and van Strijp, J.A.G. (2014). Neutrophil-mediated phagocytosis of *Staphylococcus aureus*. Front Immunol *5*, 467 DOI: 10.3389/fimmu.2014.00467.

Lewis, K. (2008). Multidrug tolerance of biofilms and persister cells. Curr Top Microbiol Immunol *322*, 107–131 DOI: 10.3389/fimmu.2014.00467.

Limoli, D.H., Jones, C.J., and Wozniak, D.J. (2015). Bacterial extracellular polysaccharides in biofilm formation and function. Microbiol Spectr *3* DOI: 10.1128/microbiolspec.MB-0011-2014.

Lovley, D.R., and Malvankar, N.S. (2015). Seeing is believing: Novel imaging techniques help clarify microbial nanowire structure and function. Environ Microbiol *17*, 2209–2215 DOI: 10.1111/1462-2920.12708.

Mittal, R., Sharma, S., Chhibber, S., and Harjai, K. (2006). Effect of macrophage secretory products on elaboration of virulence factors by planktonic and biofilm cells of *Pseudomonas aeruginosa*. Comp Immunol Microbiol Infect Dis *29*, 12–26 DOI: 10.1016/j.cimid.2005.11.002.

Morici, L.A., Carterson, A.J., Wagner, V.E., Frisk, A., Schurr, J.R., Zu Bentrup, K.H., Hassett, D.J., Iglewski, B.H., Sauer, K., and Schurr, M.J. (2007). *Pseudomonas aeruginosa* AlgR represses the Rhl quorum-sensing system in a biofilm-specific manner. J Bacteriol *189*, 7752–7764 DOI: 10.1128/JB.01797-06.

Mosser, D.M., and Edwards, J.P. (2008). Exploring the full spectrum of macrophage activation. Nat Rev Immunol *8*, 958 DOI: 10.1038/NRI2448.

Mulcahy, H., Charron-Mazenod, L., and Lewenza, S. (2010). *Pseudomonas aeruginosa* produces an extracellular deoxyribonuclease that is required for utilization of DNA as a nutrient source. Environ Microbiol *12*, 1621–1629 DOI: 10.1111/j.1462-2920.2010.02208.x.

Neu, T.R., and Lawrence, J.R. (2015). Innovative techniques, sensors, and approaches for imaging biofilms at different scales. Trends Microbiol *23*, 233–242 DOI: 10.1016/j.tim.2014.12.010.

Nilsson, M., Rybtke, M., Givskov, M., Høiby, N., Twetman, S., and Tolker-Nielsen, T. (2016). The dlt genes play a role in antimicrobial tolerance of *Streptococcus mutans* biofilms. Int J Antimicrob Agents *48*, 298–304 DOI: 10.1016/j.ijantimicag.2016.06.019.

Otto, M., (2009). *Staphylococcus epidermidis*—the' accidental' pathogen. Nat Rev Microbiol *7*, 555 DOI: 10.1038/nrmicro2182.

Pinchuk, G.E., Ammons, C., Culley, D.E., Li, S.M.W., McLean, J.S., Romine, M.F., Nealson, K.H., Fredrickson, J.K., and Beliaev, A.S. (2008). Utilization of DNA as a sole source of phosphorus, carbon, and energy by *Shewanella* spp.: Ecological and physiological implications for dissimilatory metal reduction. Appl Environ Microbiol *74*, 1198–1208 DOI: 10.1128/AEM.02026-07.

Prabhakara, R., Harro, J.M., Leid, J.G., Harris, M., and Shirtliff, M.E. (2011). Murine immune response to a chronic *Staphylococcus aureus* biofilm infection. Infect Immun *79*, 1789–1796 DOI: 10.1128/IAI.01386-10.

Prigent-Combaret, C., Vidal, O., Dorel, C., and Lejeune, P. (1999). Abiotic surface sensing and biofilm-dependent regulation of gene expression in *Escherichia coli*. J Bacteriol *181*, 5993–6002.

Scherr, T.D., Hanke, M.L., Huang, O., James, D.B.A., Horswill, A.R., Bayles, K.W., Fey, P.D., Torres, V.J., and Kielian, T. (2015). *Staphylococcus aureus* biofilms induce macrophage dysfunction through leukocidin AB and alpha-toxin. MBio *6*, e01021-15 DOI: 10.1128/mBio.01021-15.

Schommer, N.N., Christner, M., Hentschke, M., Ruckdeschel, K., Aepfelbacher, M., and Rohde, H. (2011). *Staphylococcus epidermidis* uses distinct mechanisms of biofilm formation to interfere with phagocytosis and activation of mouse macrophage-like cells 774A. 1. Infect Immun *79*, 2267–2276 DOI: 10.1128/IAI.01142-10.

Serra, D.O., Richter, A.M., and Hengge, R. (2013). Cellulose as an architectural element in spatially structured escherichia coli biofilms. J Bacteriol *195*, 5540–5554 DOI: 10.1128/JB.00946-13.

Shi, C., and Pamer, E.G. (2011). Monocyte recruitment during infection and inflammation. Nat Rev Immunol *11*, 762 DOI: 10.1038/nri3070.

Shirtliff, M., and Leid, J.G. (2009). The role of biofilms in device-related infections (Springer). DOI: 10.1007/978-3-540-68119-9.

Stephen, T.L., Groneck, L., and Kalka-Moll, W.M. (2010). The modulation of adaptive immune responses by bacterial zwitterionic polysaccharides. Int J Microbiol *2010* DOI: 10.1155/2010/917075.

Summers, C., Rankin, S.M., Condliffe, A.M., Singh, N., Peters, A.M., and Chilvers, E.R. (2010). Neutrophil kinetics in health and disease. Trends Immunol *31*, 318–324 DOI: 10.1016/j.it.2010.05.006.

Tseng, B.S., Zhang, W., Harrison, J.J., Quach, T.P., Song, J.L., Penterman, J. Singh, P.K., Chopp, D.L., Packman, A.I., and Parsek, M.R. (2013). The extracellular matrix protects *Pseudomonas aeruginosa* biofilms by limiting the penetration of tobramycin. Environ Microbiol *15*, 2865–2878 DOI 10.1111/1462-2920.

Vert, M.; Doi, Y.; Hellwich, K.; Hess, M.; Hodge, P.; Kubisa, P.; Rinaudo, M. Schue, F. (2012). Terminology for biorelated polymers and applications (IUPAC Recommendations 2012). Pure Appl Chem *84*, 377–410 DOI 10.1351/PAC-REC-10-12-04.

De Vos, W.M. (2015). Microbial biofilms and the human intestinal microbiome Npj Biofilms Microbiomes *1* DOI: 10.1038/npjbiofilms.2015.5.

Walker, T.S., Tomlin, K.L., Worthen, G.S., Poch, K.R., Lieber, J.G., Saavedra M.T., Fessler, M.B., Malcolm, K.C., Vasil, M.L., and Nick, J.A. (2005). Enhanced *Pseudomonas aeruginosa* biofilm development mediated by human neutrophils. Infect Immun *73*, 3693–3701 DOI: 10.1128/IAI.

73.6.3693-3701.2005.

Watnick, P.I., and Kolter, R. (1999). Steps in the development of a *Vibrio cholerae* El Tor biofilm. Mol Microbiol *34*, 586–595.

Webb, J.S., Thompson, L.S., James, S., Charlton, T., Tolker-Nielsen, T., Koch, B., Givskov, M., and Kjelleberg, S. (2003). Cell death in *Pseudomonas aeruginosa* biofilm development. J Bacteriol *185*, 4585–4592 DOI: 10.1128/JB.185.15.4585-4592.2003.

Webb, J.S., Lau, M., and Kjelleberg, S. (2004). Bacteriophage and phenotypic variation in *Pseudomonas aeruginosa* biofilm development. J Bacteriol *186*, 8066–8073 DOI: 10.1128/JB.186.23.8066-8073.2004.

Wilking, J.N., Zaburdaev, V., De Volder, M., Losick, R., Brenner, M.P., and Weitz, D.A. (2013). Liquid transport facilitated by channels in *Bacillus subtilis* biofilms. Proc Natl Acad Sci *110*, 848–852 DOI: 10.1073/pnas.1216376110.

Witko-Sarsat, V., Rieu, P., Descamps-Latscha, B., Lesavre, P., and Halbwachs-Mecarelli, L. (2000). Neutrophils: Molecules, Functions and Pathophysiological Aspects. Lab Investig *80*, 617–653 DOI: 10.1038/labinvest.3780067.

Application of Biofilms in Electricity Generation, Wastewater Treatment and Bioremediation

6

Akash Mitra and Arindam Mitra*

Department of Microbiology, School of Life Science and Biotechnology, Adamas University, Kolkata – 700126, West Bengal, India

*Corresponding author: arindam.mitra@adamasuniversity.ac.in

DOI: https://doi.org/10.21775/9781912530328.06

Abstract

Bacteria residing in biofilm mode are considered detrimental due to their enhanced ability of resistance towards chemicals, antibiotics and immune defenses but biofilms can be harnessed for beneficial processes such as wastewater treatment, bioremediation and electricity generation. In wastewater reactor systems, biofilms are grown and immobilized to enhance the rate of degradation of contaminants in wastewater and can be advantageous to a number of removal mechanisms such as biodegradation and biomineralization. Biofilm based bioremediation utilizes metabolic capacity of microbes in biofilms to degrade harmful pollutants and the process is considered cost-effective, reliable and widely used in industry. In microbial fuel cells, biofilms are deposited on electroconductive surfaces such as anode and cathode and during electron transfer, chemical energy is converted into usable electrical energy which can then be harvested by appropriate electronic devices. Microbial Fuel Cells are a promising sustainable technology that can reduce the dependency on fossil fuels and will be used more in the future. Here we discuss recent studies on applications of biofilms in microbial fuel cells, wastewater treatment and bioremediation.

Microbial Fuel Cells

Energy production has been one of the greatest concerns in the twenty first century. The chemical organic fuels that are used for energy generation are typically non-renewable resources and would be likely depleted or exhausted in the near future. The replacement of fossil fuels requires the use of numerous energy sources and energy carriers to meet the demands of different end uses. In contrast, Microbial Fuel Cells (MFC) can convert chemical energy into electricity via a chemical reaction with oxygen or another oxidizing agent. Biofilms can be incorporated into the MFCs which can degrade organic substances and produce electricity in the process of electron transfer. In fuel cells, biofilms are grown on the anode of the fuel cell and provided with organic substances as substrates for up taking of electrons and move them to the cathode and generate electricity in the process. Different types of biodegradable wastes which are generated in various industrial and household processes can be used as a substrate to generate electricity by providing these to the MFCs. It was shown that the resistance of charge transfer is reduced by about 16% via the use of biofilm-based fuel cells and exhibited improvement of power generation in a recent study using Electrochemical Impedance Spectroscopy (Baranitharan *et al.*, 2015). In a similar study, it was demonstrated that the power density increases by almost two units when the external resistance of a MFC decreases from 1000 to 50 ohm (Zhang *et al.*, 2011). Investigations are carried on to determine suitable microbial biofilms along with efficient substrates in MFCs. Another study used cellulose as an electron donor for obtaining renewable energy via MFC (Ren, Steinberg and Regan, 2008). The identification of *Pseudomonas aeruginosa* as an anode respiring bacteria was an important discovery as this bacterium is mesophilic, non-fermentative and aerobic. It was shown that *P.aeruginosa* can utilize pentose and hexose effectively through anode respiration where glucose utilization is more effective than sucrose or fructose and the study also revealed that electron donor type in the anodic compartment controls the growth of anodic biofilm in the MFC (Ali *et al.*, 2018). Not only anodic biofilms can help in the MFCs for enhanced power generation but also the use of biofilms grown on cathodes can prove to be an essential tool for the increased power generation. The biofilms grown on the cathode can act as catalysts for the oxygen reduction reactions (ORR). Another study investigated the energy production with a platinum-carbon based cathode on which biofilms were allowed to grow spontaneously. The study demonstrated that the highest power generation was achieved when aerobic biofilm act as catalyst for ORR in synergy with Platinum in biohybrid cathode (Massaglia, Fiorello, *et al.*, 2018). The application of the MFCs are diverse and have an application in remote sensors and data transmission in the marine environment. The performance of carbon felt-based anodic electrode containing MFC in laboratory and actual marine conditions as floating fuel cells have been tested. The results showed stability in both cases and fuel cells could continuously produce energy by using sea water as fuel and electrolytes (Massaglia, Margaria, *et al.*, 2018). This process is an advantage because of the abundance of sea water and can replace sea batteries in the future.

Microbes can metabolize a variety of nutrients and survive through adverse conditions. It has been previously reported that certain microbes can use urine as its feedstock and transfer electrons to an inert substance on the electrode of fuel cells. This process can enhance the generation of electrical energy and this was demonstrated. In this study two sets of 32 cells were prepared which were connected in parallel and activated sludge from a wastewater treatment plant was combined with human urine. The voltage was stabilized and a power density of $23mWm^{-2}$ with a current production of $65mAm^{-2}$ was produced for about 120 days (Cid *et al.*, 2018). Recently, a team dedicated a study to analyze an important aspect of electroactive biofilms i.e. the Extracellular Polymeric Substances (EPS) that is secreted by the biofilms. *Geobacter sulferreducens* was cultivated under MFC mode which produced a maximum current density of 172 ± 29 $\mu A\ cm^{-2}$. In this study, the EPS was extracted and fractioned and the amount of soluble EPS was found to be much greater than the tightly and loosely bound EPS and contained several proteins (Stöckl *et al.*, 2019). Constructed Wetlands have been extensively applied for industrial wastewater treatment and implementing the same with MFCs can have additional beneficial effect such as increased treatment capabilities and greenhouse gas assimilation (Saba *et al.*, 2018). This can be also used to treat wastewater and generate electricity. One of the main concerns regarding these processes are whether the microbes in the biofilm condition transfers electrons directly to the electrode or uses an intermediate molecular player to carry out the process. Moreover, presence of certain compounds such as nitrate causes performance instability and hinders with the functional biofilm in the MFCs. A novel strain of *Mycobacterium* sp. was isolated from an MFC fed with nitrate medium and inoculated with a sediment from a heavily polluted river. The isolate could transfer electron to the anode directly and it could denitrify the nitrate and simultaneously produce electricity (Eb- *et al.*, 2018). This is an example of dual advantage of enhanced power generation and wastewater treatment.

Since substrates are an important consideration of MFC, it is very important to harness substrates for the biofilms in the fuel cell with cost effectiveness. Cashew apple is a low-cost substrate which generated an open circuit voltage of 0.4 V, maximum power density of 31.57 mW/m^2 and a current density of 350 mA/m^2 (Divya Priya and Pydi Setty, 2019). Cellulose, pentose and hexose have also been demonstrated as suitable substrates for the biofilms in the MFC (Ren, Steinberg and Regan, 2008; Ali *et al.*, 2018). Another recent study demonstrated that Rice bran coupled with pond bottom mud can have an enhanced electricity generation which is not possible if the mentioned two are separately used as substrates for the MFCs. Phylogenetic analyses revealed the mutualism of the species of *Bacteriodes, Clostridium* and *Geobacter* in the consortium and a total of about 26% increase in electricity production was observed (Yoshimura *et al.*, 2018). MFCs can be also used in the BOD assessment which is also a part of wastewater treatment. Like remote sensors and data transmission systems (Massaglia, Margaria, *et al.*, 2018), MFCs are also used as biosensors for detecting Biochemical Oxygen Demand (BOD) and biotoxicity. In a recent study it was shown that an MFC containing pure culture of *Shewanella lohica* PV-4 was much more sensitive under toxic shock compared to a mixed culture consortium and therefore delivered higher sensitivity for biotoxicity monitoring (Yi *et al.*, 2018). A similar study was performed for detecting the biotoxicity of carbon monoxide (CO) on the anodic biofilm produced from the consortium in domestic wastewater. A proportionate relationship was observed between the increase of CO concentration with the voltage drop over a time span of 1 hour (Zhou *et al.*, 2018). The biosensing of toxic agents through a single electrode does provide a picture of the toxic shock, however the combined toxic shock created by both organic matter and toxic agents remains apparent. To address this, Zhao *et.* al have shown that sequential flowing membrane-less microbial fuel cells can detect the combined shock of organic matter coupled with toxic agents (Zhao *et al.*, 2019). The biofilm on both the anode and cathode contributes to the electricity generation and makes the effluent come in contact simultaneously with both the electrodes. This could result in a better sensing of the toxic shock agents since both would be interfered in the process. Apart from toxicity sensing, MFCs can be applied as a biosensor for various anaerobic processes such as anaerobic digestion. A linear relationship between current density and acetate (as sole carbon source) was successfully established in a double-chambered MFC from anaerobic digestion of effluent (Sun *et al.*, 2019).

A lot of possible approaches have been undertaken to increase the power generation by MFCs. An effective way is to directly transfer the electrons from the biofilms to the anodic electrode. Low intensity magnetic field can reduce the startup time and cause a significant increase in the power density and electricity generation (Li, Wang and Liu, 2018). Short term application of magnetic field has been shown to increase the conductance of anodic biofilms (Li, Lesnik and Liu, 2017). The distance of electron transfer depends on the type of biofilm on the electrode as well. Mixed species biofilms tend to have inter-species interactions that makes the consortium transport electron to greater distances of about 1mm which is tenfold of that observed in case of single species biofilm (Li *et al.*, 2016). Microbes in the biofilms may also produce certain pigments which can enhance the electron flow. Pyocyanin, a pigment developed by *Pseudomonas aeruginosa* PAO1 acts as an extracellular shuttle (Ren *et al.*, 2018). Cyclic voltammetry and Surface enhanced Raman spectroscopy revealed light driven rapid bio-electron transfer.

Not only nutrients such as glucose are used as substrates but also macroalgal species such as *Laminaria digitata* has proven to be an effective biomass substrate for MFCs, although the pathway would mostly tend towards glucose formation with the transfer of electrons to the electrodes (Zhao *et al.*, 2018). Efficient substrates are optimized such that both the electricity generation and biosensing capacity of the biofilm based MFCs are enhanced and relatively high throughput power generation is achieved.

Application in wastewater treatment

Pollution of aquatic environment has been a major concern over the decades. There are various categories of water pollutants and a significant pollutant, called a priority pollutant can have a maximum contamination level of 1 µg/L. Among the priority pollutants, pentachlorophenol (PCP, C_6Cl_5OH) which is used as pesticide, disinfectant and wood preservative is a widespread environmental contaminant in soils, surface water and ground water. One study demonstrated that a hydrogen-based membrane biofilm reactor successfully removed almost 10 mg/L PCP with minimal intermediate accumulation and complete denitrification of 20 mg-N/L nitrate (Long *et al.*, 2018). Biofilm producing *Xanthobacter* and *Azospira* could reductively dechlorinate PCP which is subsequently mineralized to CO_2. In another study regarding water pollution treatment, the bacterial community structure and diversity was assessed using the rotating bio-disk (RB) procedure in a semi-industrial pilot plant. The *Salmonella* community was particularly monitored and the effects of UV were studied. Along with *Salmonella* several other species such as *Enterobacter agglomerans, Cronobacter sakazakii, and Pantoea agglomerans* are also identified which have potential for bioremediation activities and all of them occurred in biofilm samples (Yousra Turki *et al.*, 2017). Although biological denitrification is a non-substitutable municipal wastewater treatment for nitrate removal, it is not a very cost-effective process because of the additions of external carbon source. The synergy of electrochemical reactions and biological reaction in bio-electrochemical systems have enhanced the electrochemical reduction of N_2O, which in turn, enhances the nitrate removal process (Jiang *et al.*, 2018). Over the past ten years, the demand for aquatic products have increased to a huge level as the demand for food fish supply has increased at an annual rate of 6.2% (Otero *et al.*, 2017). Major concerns with aquaculture wastewater include removal of organic matter and nutrients, specifically suspended solids, oxygen depletion, nitrogen and phosphorus. Biofilters are used to remove nitrogen compounds from water, convert them into N_2O which is a potential greenhouse gas causing global warming. A study reported development of a synthetic lichen type biofilm combining fungus *Mucor indicus* and the microalga *Chlorella vulgaris* grown together on a polymer matrix, which together could utilize phosphorus and nitrogen compounds and convert them to proteins and other cellular products. This mycoalgae biofilm could also clean the water from nutrient pollutants as the algae are attached to the biofilm leaving purified water at the end of the process (Barnharst, Rajendran and Hu, 2018). Under 25 mg L^{-1} of total ammonia-N conditions, the biofilm could reduce it to undetectable limits within 48 h. Acetonitrile is a water-soluble toxic substance that is often discharged in industrial and laboratory wastewater. Effort has been made for creating efficient technology in wastewater treatment to reduce the discharge of acetonitrile. A study constructed a recombinant protein using nitrilase gene isolated from nitrile degrading bacterium (*Rhodococcus rhodochrous* BX2) and cloned it into a biofilm forming bacterium (*Bacillus subtilis* N4) that expressed the recombinant protein upon IPTG induction. The biofilm thus formed had strong nitrile degrading capability and completely degraded the initial concentration of acetonitrile (800 mg L^{-1}) within 24 h in a moving bed biofilm reactor (Zhang *et al.*, 2016).

Animals in aquaculture are very easily infected by pathogens and tetracycline antibiotics containing oxytetracycline (OTC) are used to minimize infections. It has been observed that 75% of OTC is released into the aquatic environment. To biodegrade OTC, a study was performed to characterize the biodegradation of oxytetracycline by the strain *Ochrobactrum* sp. KSS10 under varying environmental conditions such as temperature, carbon source etc. About 63% of OTC degradation along with 98% of nitrate removal were obtained within 96 hours (Shao *et al.*, 2019). A similar experiment was performed with samples of influent and effluent sewerage of a wastewater treatment plan in Slovakia and predominant antibiotics that were found in the influent was azithromycin, clarithromycin and ciprofloxacin. Biofilms from the wastewater influent and effluent contained antibiotic resistant *Enterococcus* spp. which were tested to be resistant to lower doses of ampicillin and ciprofloxacin and coliform bacteria which were observed to be resistant to ampicillin and gentamicin. The removal efficiency of these antibiotics, notably azithromycin was 30% in most cases (Jiang *et al.*, 2018).

Formaldehyde is a volatile organic pollutant present in wastewater and is categorized as carcinogenic. In a recent study, a new method with a membrane-aerated biofilm reactor was successfully devised for treatment of formaldehyde in wastewater and an efficient degradation was obtained without generation of any secondary pollution. For about 3 kg HCHO/m^3, around 97% removal rate was achieved (Jiang *et al.*, 2018). Onsite wastewater treatment systems have been shown to contain a critical amount of micropollutants such as per- and polyfluoroalkyl substances (PFAS) and a non-expensive purification system is generally lacking for these pollutants. A study attempted to remove these PFAS using biochar filters with no biofilm and compared with biochar filters with activated and inactivated biofilm. While the former one was highly efficient in removal of PFAS with the range of C_6 to C_8 , the latter one could efficiently remove PFAS of the range of C_{11} which are proteinophilic through surface adsorption (Dalahmeh, Alziq and Ahrens, 2019). A very similar cause of water pollution is the disposal of the pharmaceutically active compounds (PhACs) in the water bodies which include various drugs, antibiotics etc. A method to remove these PhACs was devised by using a combination of biological activated carbon (BAC) filter and ultrafiltration (UF). The idea was to evaluate the biological activity of the biofilm that developed on the activated carbon particles in two different phases, one with 9200 bed volumes and other with bed volumes between 9,200 and 13,800, were observed and characterized. Although physical adsorption of PhAC was the major removal mechanism, a significant amount of biofilm activity removed 22% of ciprofloxacin and 32% of azithromycin (Sbardella *et al.*, 2018). Oil spills are one of the major causes of contamination of water bodies. From two oil reservoirs, a team of scientists isolated nine thermophilic bacterial strains belonging to *Bacillus* and *Geobacillus* from crude oil located in the Cauvery river basin, Karaikal, India. Their biodegradation efficiency of hydrocarbons were found to be quite impressive with the highest being 94% for the strain IR4 and screening for enzyme activity, it was found that the alkane hydroxylase had the highest activity (Elumalai *et al.*, 2019). A similar study concerning petroleum hydrocarbon pollutants from wastewater was addressed with serial moving bed biofilm reactor by connecting two moving bed biofilm reactors for the purpose. Optimizing the influent concentration of petroleum hydrocarbon along with the retention time created an improved removal efficiency of petroleum pollutants (Qaderi, Sayahzadeh and Azizi, 2018). Aquatic plants such as *Myriophyllum aquaticum* has been used to treat wastewater and it exhibits high potential for nitrogen and phosphorus removal (Liu *et al.*, 2016). Incorporation of activated sludge and formation of biofilm synchronously with the growth of the plant can enhance the COD removal along with nitrogen and phosphorus removal relative to those system with only the presence of the plant (Guo *et al.*, 2019). A study was conducted to observe the effects of the reactor wall attached microalgal biofilm on the nutrient removal performance from wastewater. Two reactors were set up where one contained attached biofilm and the other did not. The reactor with biofilm had statistically higher removal efficiencies of phosphorus and nitrogen (Zhang *et al.*, 2016). Cotreatment of hydraulic fracturing produced water (PW) with municipal wastewater in a laboratory scale moving bed biofilm reactor resulted in removal rate of nitrogen above 60% when the mixing ratio slowly reached maximum. Shotgun metagenomic analysis revealed the functional diversity of nitrite-oxidizing bacteria. Since shortage of water sources has always been a concern, there had been numerous approaches to make water potable. Recycling and decontaminating secondary effluent have been taken into consideration to address this, however decontaminating the trace organic contaminants and high concentration of organic matter has been a challenge. UV-based processes have been very efficient in removing the trace organic contaminants and combining the UV-based process along with biofilm-based removal had a much significant reduction of organic contaminants along with dissolved organic carbon. Hence, this can be an effective wastewater reclamation process with an effective treatment (Yan *et al.*, 2019). A worldwide application of nanotechnology has led to enormous release of nanoparticles in wastewater treatment systems and has a potential to impair the disposal of sewage sludge. Extracellular polymeric substances (EPS) of biofilms in anaerobic granular sludge and activated sludge can be used as a highly efficient adsorbing material for the nano-ZnO and nano-CuO (Yan *et al.*, 2019). Retention of the advantageous biofilm inside the reactor is one of the major concerns for wastewater treatment for its prolonged activity. Along with retention, the biovolume of the biofilm is also important. *Nitrosomonas europaea,* an ammonia oxidizing bacteria reduces reactive nitrogen in wastewater treatment before the water is discharged into water bodies. A recent study revealed that introduction of *N. europaea* in co-culture with heterophilic *Pseudomonas aeruginosa* in the reactor causes a substantial increase in the biovolume and the retention capacity of the biofilm increased in a dual species biofilm and also enhanced the nitrogen removal process (Petrovich *et al.*, 2017). Generally, biofilms developed for wastewater treatment detach themselves from the carrier media once the available thickness exceeds 2 mm. A recent study compared the feasibility for wastewater treatment with and without basalt fibers as carrier media. Basalt fiber media can form aggregate of up to 10cm. FTIR and confocal laser scanning microscopy analyses showed that both reactors had removed about 90% of the chemical oxygen demand (COD) but only the reactor with the basalt fiber carrier media achieved high denitrification abilities of around 60-80% (Qaderi, Sayahzadeh and Azizi, 2018). In a similar study the conventional basalt fiber carrier was compared with a modified version of the same where the modified version of the carrier had hydrophilic cationic polymer (hexadecyl trimethyl ammonium chloride, CTAC). This modified basalt fiber had a more hydrophilic and positively charged surface. The bio-affinity of both the carrier was evaluated by comparing *Bacillus subtilis* adhesion and activated sludge immobilization tests. The result showed that modified carriers had higher biomass attachment capacity with shorter biofilm formation time. Upon FTIR analysis it was seen that more bacteria adhered to the modified carrier surface due to electrostatic and hydrophilic attractive interactions in the initial period and EPS further aggregated their adhesion. The immobilization ratio increased from 141.64% (unmodified) to 224.74% (modified) (Liu *et al.*, 2017). In another study for effective carriers, in 2016 natural ventilation trickling filters (NVTF) with three different carriers were used to treat domestic wastewater. The carriers used were sponge, zeolite and ceramsite. All three showed table capability of COD removal and ammonia nitrification. It was found that greater biomass was attached to sponge than the other two. Sponge carrier had the most effective COD removal with ceramsite being second. For nitrogen removal, sponge carrier again was the most effective and zeolite came second (Zhang *et al.*, 2016).

Existing conventional attached growth wastewater systems are greatly hampered by long start-up times and erosion. I. Namabi *et al.* developed an electrically bound biofilm reactor consisting of a nematic liquid crystal electrode (NLCE) as the anode obtained from an inutile desktop monitor and evaluated Al, Cu and Pt as candidates for cathode. Optimizing the condition to 1 V and using the Pt-NLCE electrode enhanced biofilm attachment and reduced suspended solids up to 79.3% and a biodegradation rate of 71.2% was observed. High biofilm stability and rapid start-up times were also observed. Hence, enhanced biofilm attachment combined with rapid start-up times can be very efficient for approaches of wastewater treatment efforts (Srinivasan, Nambi and Senthilnathan, 2019).

A significant amount of wastewater is produced in the dairy industry with a high potential to pollute the environment, if not treated properly. Anaerobic treatment technologies have been successfully applied due to its high biodegradability. A 2015 review focused on biofilm reactors for anaerobic treatment of dairy wastewater. Several kinds up-flow anaerobic filter packed with varying media has provided higher treatment efficiency and has high tolerance against heavy metal toxicity (Jiang *et al.*, 2018). Mining industries encounter challenges of treating sulfur-containing wastewater such as acid mine drainage (AMD). A study showed that revolving algal biofilm reactor had a sulfate removal efficiency of up to 46% and removal rate of up to 0.56 g/L-day, much higher than those obtained in suspension algal culture. Along with sulfate removal, the reactors also showed enhanced COD, ammonia and phosphorus removal. This is a potentially useful technology which can also be implemented in case of municipal wastewater treatment facilities (Jiang *et al.*, 2018).

Application in bioremediation

The major challenge with most of the strategies for reducing environmental pollutants is high costs and poor efficiency. In a scenario like this, bioremediation has proved to be a cost effective for cleaning pollutants and maintaining its eco-friendliness. In recent times, biofilms have gained attention for bioremediation. Quorum sensing (QS) is one of the several mechanisms which is used by bacteria in biofilms for signaling purposes. It also plays an affirmative role in the degradation of polycyclic aromatic hydrocarbons (PAH). In a study of the marine bacterium *P. aeruginosa* N6P6 the QS genes (*lasI* and *rhlI)* had a significant increase in their expression with an increased PAH degradation (Mangwani, Kumari and Das, 2015). Hence, quorum sensing has a significant role in regulating the genes involved in bioremediation process. Moreover, the inherent stress-resistant and metabolic dynamic qualities of biofilm make its use in bioremediation of assorted industrial and environmental pollutants (Mangwani, Kumari and Das, 2017). In another study by the same group of scientists, degradation of phenanthrene and pyrene was evaluated with evaluation of certain parameters like solubility of PAHs, EPS and catechol 2,3-deoxygenase (C23O) activity. *P. aeruginosa* N6P6 and *P. pseudoalcaligenes* NP103 showed chemotactic movements towards both the PAHs. The solubility of the PAHs increased with an increase in the EPS concentration and about 70% phenanthrene and 55% pyrene degradation was observed. A significant increase in the C23O was also recorded (Mangwani, Kumari and Das, 2017). In a similar study for the bioremediation of phenanthrene, application of clay minerals such as calcinated bentonite (CB) and calcinated organobentonite (COB) was evaluated. Both of them showed high efficiency of phenanthrene removal. Diverse bacteria taxa were isolated from both CB and COB. COB contained 18% and 21% of *Sphingomonadaceae* and *Rhodobacteraceae* and on the other hand CB had 23% of *Spingomonadaceae.* The removal rate in COB was around 99% within 2 days and 95% for CB within 3 days (Huang Zhang and Zhu, 2013). Apart from PAH, benzene, toluene, ethylbenzene and xylene (BTEX) are considered dangerous volatile hydrocarbons with high carcinogenic impacts. BTEX can penetrate groundwater and contaminate the soil gradients. A major way to tackle this problem is to inject oxygen releasing compounds in the subsurface zones of the soil where these compounds decompose to oxygen and hydroxyl radicals by exposure to water and this accelerates contaminant degradation (Mosmeri *et al.*, 2019). In a recent study the utilization of CaO_2 in bioremediation was investigated for benzene removal in continuous flow sand-packed columns. Since calcium peroxide supplies oxygen adequately, there was a logarithmic increase in the number of planktonic bacteria and gradually 100% benzene removal was achieved by the end of the experiment. It was also observed that the diversity of microbes decreased with the increase in CaO_2 injection but on the other hand benzene degrading species such as *Silanimonas, Arthrobacter* and *Pseudomonas* spp. dominated (Mosmeri *et al.*, 2019).

Tellurite is one of the toxic substances that finds its way into the environment causing toxicity. Its lethal concentration is several folds lower than that of other metals like Fe, Hg and Zn. Some microbes have developed resistance against tellurite and can reduce it to elemental tellurium. A study was conducted in 2018 and tellurite reducing strain was isolated from Zuari estuary, Goa, India. The strain was identified as *Shewanella baltica* which could tolerate 5.5 mM potassium tellurite with an MIC of 6 mM. It was found that the strain could completely reduce 2 mM tellurite to elemental tellurium during the late stationary phase. Interestingly, the strain initiated the biosynthesis of Te-nanoparticles which peaked at its mid-log phase (Vaigankar *et al.*, 2018). Apart from tellurite, another source of environmental hazard is arsenic (As). Arsenic pollution mainly occurs by volcanic activities but a part of it is also due to consequence of human activities such as processing of geological materials such as coal and metalliferous ores. A long-term exposure to arsenic can cause critical effects on human health. According to the World Health Organization (WHO), 10 µg As/L in water is the limit for As in drinking water. In the High Altitude Andean Lakes (HAALs) which are shallow lakes situated in Argentina, Chile, and Bolivia are prone to active volcanism and has an arsenic concentration ranging between 800 µg As/L and 33,800 µg As/L (Zannier *et al.*, 2019). Another study attempted to characterize the biofilm from the HAALs found two strains of particular importance. These were *Acinetobacter* sp. (Ver 3) and *Exiguobactrium* sp. (S17). These strains could tolerate high arsenic concentration than their phylogenetically related strains from German collection of microorganisms and cell cultures. Hence, these strains can be used for potential bioremediation of arsenic from the environment (Zannier *et al.*, 2019).

In recent years, various kinds of microbial electrochemical systems (MES) have been developed as remedy for soil contamination. The MES is similar to a microbial fuel cell and typically consists of an anode, a cathode and a separator (Wang and Ren, 2013). Biodegradable substrates are oxidized and the electrons are released at the anode. The electrons travel via external electrical circuit to the electron acceptor at the cathode. Sometimes, the terminal acceptor for electrons is low in its redox potential and therefore additional power supply is required to drive the overall reaction. MES is mostly used in remediation of petroleum contaminants. A recent review on MES application stated that a removal efficiency of about 89% (approx.) for petroleum hydrocarbon was achieved in a pilot scale after 120-day treatment (Wu *et al.*, 2018). Apart from organic contaminants, heavy metals are one of the persistent contaminants that can be remediated using MES. A study applied soil MES in heavy metal bioremediation, and observed that Copper, Cadmium and Lead could migrate and got enriched in the cathode region and after 143 and 108 day operation, 3% of Cadmium and 44% of Lead respectively were removed from the anode region (Wang *et al.*, 2016).

Concluding remarks and future trends

Use of bacterial biofilms in Microbial Fuel Cells is one the new age technology which not only generates electricity but also have other applications including wastewater treatment and bioremediation. MFC reduces the use of fossil fuels and thereby have a beneficial effect on climate and environment. MFC also generate lesser secondary pollutant and are cost effective. Studying the function of biofilms in electron transfer also can shed light on new strategy to optimize the performance of Microbial Fuel Cells. Surface alterations and genetic changes can shed light on how to optimize fuel cells. Besides modification of any steps in biofilm formation can further open up new avenues to maximize performance of Microbial Fuel Cells. Choice of microbes, substrates and media for bacteria can also impact the performance of biofilms. Besides, biofilms have applications in wastewater treatment and bioremediation, two important areas where recombinant technology and metagenomics can bring a major improvement in the field. Another area is to unravel the exact basis of community formation from different species (mixed species biofilm) and how that population of microbes can optimize the performance.

References

Ali, N. *et al.* (2018) 'Characterization of the Electric Current Generation Potential of the Pseudomonas aeruginosa Using Glucose, Fructose, and Sucrose in Double Chamber Microbial Fuel Cell', *Iranian Journal of Biotechnology*, 15(4), pp. 216–223. doi: 10.15171/ijb.1608.

Baranitharan, E. *et al.* (2015) 'Effect of biofilm formation on the performance of microbial fuel cell for the treatment of palm oil mill effluent', *Bioprocess and Biosystems Engineering*, 38(1), pp. 15–24. doi: 10.1007/s00449-014-1239-9.

Barnharst, T., Rajendran, A. and Hu, B. (2018) 'International Biodeterioration & Biodegradation Bioremediation of synthetic intensive aquaculture wastewater by a novel feed-grade composite bio fi lm', 126(August 2017), pp. 131–142. doi: 10.1016/j.ibiod.2017.10.007.

Cid, C. A. *et al.* (2018) 'Urine microbial fuel cells in a semi-controlled environment for onsite urine pre-treatment and electricity production', *Journal of Power Sources*, 400(June), pp. 441–448. doi: 10.1016/j.jpowsour.2018.08.051.

Dalahmeh, S. S., Alziq, N. and Ahrens, L. (2019) 'Potential of biochar filters for onsite wastewater treatment: Effects of active and inactive biofilms on adsorption of per- and polyfluoroalkyl substances in laboratory column experiments', *Environmental Pollution*. Elsevier Ltd, 247, pp. 155–164. doi: 10.1016/j.envpol.2019.01.032.

Divya Priya, A. and Pydi Setty, Y. (2019) 'Cashew apple juice as substrate for microbial fuel cell', *Fuel*. Elsevier, 246(April 2018), pp. 75–78. doi: 10.1016/j.fuel.2019.02.100.

Eb-, E. S. *et al.* (2018) 'Enhancing the Electricity Generation and Nitrate Removal of Microbial Fuel Cells With a Novel Denitrifying', 9(November), pp. 1–10. doi: 10.3389/fmicb.2018.02633.

Elumalai, P. *et al.* (2019) 'Role of thermophilic bacteria (Bacillus and Geobacillus) on crude oil degradation and biocorrosion in oil reservoir environment', *3 Biotech*. Springer International Publishing, 0(0), p. 0. doi: 10.1007/s13205-019-1604-0.

Guo, X. *et al.* (2019) 'A wastewater treatment system combining Myriophyllum aquaticum and activated sludge: Optimization of construction conditions and evaluation of wastewater treatment performance', *Journal of Environmental*

Management. Elsevier, 235(January), pp. 105–111. doi: 10.1016/j.jenvman.2019.01.032.

Huang, Y., Zhang, J. and Zhu, L. (2013) 'Evaluation of the application potential of bentonites in phenanthrene bioremediation by characterizing the biofilm community', *Bioresource Technology*, 134, pp. 17–23. doi: 10.1016/j.biortech.2013.02.009.

Jiang, X. *et al.* (2018) 'Electrochemical study of enhanced nitrate removal in wastewater treatment using biofilm electrode', *Bioresource Technology*, 252, pp. 134–142. doi: 10.1016/j.biortech.2017.12.078.

Li, C. *et al.* (2016) 'Millimeter scale electron conduction through exoelectrogenic mixed species biofilms', *FEMS Microbiology Letters*, 363(15), pp. 1–6. doi: 10.1093/femsle/fnw153.

Li, C., Lesnik, K. L. and Liu, H. (2017) 'AC PT US CR', *Biotechnology Advances*. Elsevier Inc. doi: 10.1016/j.biotechadv.2017.07.010.

Li, C., Wang, L. and Liu, H. (2018) 'Enhanced redox conductivity and enriched Geobacteraceae of exoelectrogenic biofilms in response to static magnetic field', *Applied Microbiology and Biotechnology*, 102(17), pp. 7611–7621. doi: 10.1007/s00253-018-9158-3.

Liu, F. *et al.* (2016) 'Nitrogen removal and mass balance in newly-formed Myriophyllum aquaticum mesocosm during a single 28-day incubation with swine wastewater treatment', *Journal of Environmental Management*, 166, pp. 596–604. doi: 10.1016/j.jenvman.2015.11.020.

Liu, R. *et al.* (2017) 'Can biofilm affect alum sludge adsorption: An engineering scope in a novel biofilm reactor for wastewater treatment', *Chemical Engineering Journal*, 328, pp. 683–690. doi: 10.1016/j.cej.2017.07.081.

Long, M. *et al.* (2018) 'SC', *Water Research*. Elsevier Ltd. doi: 10.1016/j.watres.2018.06.071.

Mangwani, N., Kumari, S. and Das, S. (2015) 'Involvement of quorum sensing genes in biofilm development and degradation of polycyclic aromatic hydrocarbons by a marine bacterium Pseudomonas aeruginosa N6P6', *Applied Microbiology and Biotechnology*, 99(23), pp. 10283–10297. doi: 10.1007/s00253-015-6868-7.

Mangwani, N., Kumari, S. and Das, S. (2017) 'Bacterial biofilms and quorum sensing: fidelity in bioremediation technology', *Biotechnology and Genetic Engineering Reviews*, 32(1–2), pp. 43–73. doi: 10.1080/02648725.2016.1196554.

Mangwani, N., Kumari, S. and Das, S. (2017) 'Marine Bacterial Biofilms in Bioremediation of Polycyclic Aromatic Hydrocarbons (PAHs) Under Terrestrial Condition in a Soil Microcosm', *Pedosphere*. Soil Science Society of China, 27(3), pp. 548–558. doi: 10.1016/S1002-0160(17)60350-3.

Massaglia, G., Fiorello, I., *et al.* (2018) 'Biohybrid Cathode in Single Chamber Microbial Fuel Cell', *Nanomaterials*, 9(1), p. 36. doi: 10.3390/nano9010036.

Massaglia, G., Margaria, V., *et al.* (2018) 'In situ continuous current production from marine floating microbial fuel cells', *Applied Energy*. Elsevier, 230(June), pp. 78–85. doi: 10.1016/j.apenergy.2018.08.061.

Mosmeri, H. *et al.* (2019) 'Bioremediation of benzene-contaminated groundwater by calcium peroxide (CaO 2) nanoparticles: Continuous-flow and biodiversity studies', *Journal of Hazardous Materials*. Elsevier B.V., pp. 183–190. doi: 10.1016/j.jhazmat.2019.02.071.

Otero, M. *et al.* (2017) 'Chemosphere Removal of tricaine methanesulfonate from aquaculture wastewater by adsorption onto pyrolysed paper mill sludge', 168, pp. 139–146. doi: 10.1016/j.chemosphere.2016.10.045.

Petrovich, M. *et al.* (2017) 'Nitrosomonas europaea biofilm formation is enhanced by Pseudomonas aeruginosa', *FEMS Microbiology Ecology*, 93(5), pp. 1–25. doi: 10.1093/femsec/fix047.

Qaderi, F., Sayahzadeh, A. H. and Azizi, M. (2018) 'Efficiency optimization of petroleum wastewater treatment by using of serial moving bed biofilm reactors', *Journal of Cleaner Production*. Elsevier B.V., 192, pp. 665–677. doi: 10.1016/j.jclepro.2018.04.257.

Ren, G. *et al.* (2018) 'PT NU SC', *Bioelectrochemistry*. Elsevier B.V, p. #pagerange#. doi: 10.1016/j.bioelechem.2018.06.003.

Ren, Z., Steinberg, L. M. and Regan, J. M. (2008) 'Electricity production and microbial biofilm characterization in cellulose-fed microbial fuel cells', *Water Science and Technology*, 58(3), pp. 617–622. doi: 10.2166/wst.2008.431.

Saba, B. *et al.* (2018) 'PT US', *Bioelectrochemistry*. Elsevier B.V, p. #pagerange#. doi: 10.1016/j.bioelechem.2018.12.005.

Sbardella, L. *et al.* (2018) 'Advanced biological activated carbon filter for removing pharmaceutically active compounds from treated wastewater', *Science of the Total Environment*. The Authors, 636, pp. 519–529. doi: 10.1016/j.scitotenv.2018.04.214.

Shao, S. *et al.* (2019) 'Ecotoxicology and Environmental Safety Action of oxytetracycline (OTC) degrading bacterium and its application in Moving Bed Biofilm Reactor (MBBR) for aquaculture wastewater pre- treatment', *Ecotoxicology and Environmental Safety*. Elsevier Inc., 171(December 2018), pp. 833–842. doi: 10.1016/j.ecoenv.2019.01.040.

Srinivasan, R., Nambi, I. M. and Senthilnathan, J. (2019) 'Liquid crystal display electrode assisted bio-reactor for highly stable and enhanced biofilm attachment for wastewater treatment – A sustainable approach for e-waste management', *Chemical Engineering Journal*, 358, pp. 1012–1021. doi: 10.1016/j.cej.2018.09.082.

Stöckl, M. *et al.* (2019) 'Energy , Environmental , and Catalysis Applications Extracellular polymeric substances from Geobacter sulfurreducens biofilms in microbial fuel cells'. doi: 10.1021/acsami.8b14340.

Sun, H. *et al.* (2019) 'Innovative operation of microbial fuel cell-based biosensor for selective monitoring of acetate during anaerobic digestion', *Science of the Total Environment*. Elsevier B.V., 655, pp. 1439–1447. doi: 10.1016/j.scitotenv.2018.11.336.

Vaigankar, D. C. *et al.* (2018) 'Tellurite biotransformation and detoxification by Shewanella baltica with simultaneous synthesis of tellurium nanorods exhibiting photo-catalytic and anti-biofilm activity', *Ecotoxicology and Environmental Safety*. Elsevier Inc., 165(May), pp. 516–526. doi: 10.1016/j.ecoenv.2018.08.111.

Wang, H. *et al.* (2016) 'New process for copper migration by bioelectricity generation in soil microbial fuel cells', *Environmental Science and Pollution Research*, 23(13), pp. 13147–13154. doi: 10.1007/s11356-016-6477-8.

Wang, H. and Ren, Z. J. (2013) 'A comprehensive review of microbial electrochemical systems as a platform technology', *Biotechnology Advances*. Elsevier Inc., 31(8), pp. 1796–1807. doi: 10.1016/j.biotechadv.2013.10.001.

Wu, Y. *et al.* (2018) 'Recent advances in microbial electrochemical system for soil bioremediation', *Chemosphere*. Elsevier Ltd, 211, pp. 156–163. doi: 10.1016/j.chemosphere.2018.07.089.

Yan, P. F. *et al.* (2019) 'Efficiency of sequential UV/H 2 O 2 and biofilm process for the treatment of secondary effluent', *Environmental Science and Pollution Research*. Environmental Science and Pollution Research, 26(1), pp. 577–585. doi: 10.1007/s11356-018-3606-6.

Yi, Y. *et al.* (2018) 'Comparative analysis of microbial fuel cell based biosensors developed with a mixed culture and Shewanella loihica PV-4 and underlying biological mechanism', *Bioresource Technology*. Elsevier, 265(June), pp. 415–421. doi: 10.1016/j.biortech.2018.06.037.

Yoshimura, Y. *et al.* (2018) 'Electricity Generation from Rice Bran by a Microbial Fuel Cell and the Influence of Hydrodynamic Cavitation Pretreatment', *ACS Omega*, 3(11), pp. 15267–15271. doi: 10.1021/acsomega.8b02077.

Yousra Turki *et al.* (2017) 'Biofilms in bioremediation and wastewater treatment: characterization of bacterial community structure and diversity during seasons in municipal wastewater treatment process', *Environmental Science and Pollution Research*. Environmental Science and Pollution Research, 24(4), pp. 3519–3530. doi: 10.1007/s11356-016-8090-2.

Zannier, F. *et al.* (2019) 'Polyextremophilic Bacteria from High Altitude Andean Lakes : Arsenic Resistance Profiles and Biofilm Production', 2019.

Zhang, L. *et al.* (2011) 'Biofilm formation and electricity generation of a microbial fuel cell started up under different external resistances', *Journal of Power Sources*. Elsevier B.V., 196(15), pp. 6029–6035. doi: 10.1016/j.jpowsour.2011.04.013.

Zhang, X. *et al.* (2016) 'Biofilm characteristics in natural ventilation trickling filters (NVTFs) for municipal wastewater treatment: Comparison of three kinds of biofilm carriers', *Biochemical Engineering Journal*. Elsevier B.V., 106, pp. 87–96. doi: 10.1016/j.bej.2015.11.009.

Zhao, N. *et al.* (2018) 'Electricity generation and microbial communities in microbial fuel cell powered by macroalgal biomass', *Bioelectrochemistry*. Elsevier B.V., 123, pp. 145–149. doi: 10.1016/j.bioelechem.2018.05.002.

Zhao, T. *et al.* (2019) 'Sequential flowing membrane-less microbial fuel cell using bioanode and biocathode as sensing elements for toxicity monitoring',

Bioresource Technology. Elsevier Ltd, pp. 276–280. doi: 10.1016/j.biortech. 2019.01.009.

Zhou, S. *et al.* (2018) 'Microbial fuel cell-based biosensor for toxic carbon monoxide monitoring', *Talanta*, 186(April), pp. 368–371. doi: 10.1016/j.talanta. 2018.04.084.

www.ingramcontent.com/pod-product-compliance
Ingram Content Group UK Ltd.
Pitfield, Milton Keynes, MK11 3LW, UK
UKHW061959041026
15591UKWH00016B/1088

9 781912 530328